MINESCAPES

Minescapes

· · · · · ·

RECLAIMING MINNESOTA'S MINED LANDS

Pete Kero

MINNESOTA
HISTORICAL
SOCIETY PRESS

The publication of this book was supported through a generous grant from the Elmer L. and Eleanor Andersen Publications Fund.

mnhspress.org

The Minnesota Historical Society Press is a member
of the Association of University Presses.

Manufactured in the United States of America.

10 9 8 7 6 5 4 3 2 1

♾ The paper used in this publication meets the minimum requirements of the American National Standard for Information Sciences—Permanence for Printed Library Materials, ANSI Z39.48-1984.

International Standard Book Number
ISBN: 978-1-68134-224-5 (paper)
ISBN: 978-1-68134-225-2 (e-book)

Library of Congress Control Number: 2022950121

The Mesabi Iron Range is located on the traditional, ancestral, and contemporary lands of Indigenous people. The Ojibwe (Anishinaabe) people, and before them the Dakota and other Native people, have cared for the land and called it home from time immemorial.

During the writing of this book, new signs were erected across the Mesabi Iron Range to help the public understand the treaties that allowed for mining and settlement of this area by non-Native people. One such sign along US Highway 169 between Chisholm and Buhl marks the boundaries of the 1854 treaty (covering northeastern Minnesota) and the 1855 treaty (covering north-central Minnesota). The lands ceded by these treaties hold great historical, spiritual, and personal significance for the Native nations and peoples of this region. By offering this land acknowledgment, I affirm tribal sovereignty of the American Indian peoples and nations in this territory and beyond.

This book, in discussing the Mesabi Iron Range from 1905 through the present, generally describes the land uses after the treaties were signed, after the original lifeways of the Native peoples who lived on this land were disrupted, and after the majority of Native peoples who originally occupied the land had been forcibly moved to tribal reservations. While I personally know very little about the Native history of this land and would not feel qualified to write about that history, the Native inhabitants certainly knew of the iron ore beneath their feet. The very name of the Biwabik Iron Formation comes from the Ojibwe *biiwaabik*, meaning iron or metal. And the name Mesabi—which is also spelled Mesaba, Missabe, and Missabay—comes from the Ojibwe *misaabe*, meaning a giant. As a result, the English word *giant* is used throughout the Mesabi Iron Range, from geological terminology (for example, the Giants Range batholith, which is the granite intrusion that serves as the range's "spine" and constitutes the Laurentian watershed divide) to place names (Giants Ridge Recreation Area) to sports teams (the Giants of Mesabi East High School). In the names used in many parts of the region, the Native heritage of this land remains evident.[1]

While this book focuses primarily on the non-Native history of mining, mineland reclamation, and mineland repurposing that occurred on this land, small portions of this story intersect with the stories of

Native people who still live in this region. For example, during my interview for chapter 3, Dave Youngman remembered Native people visiting the property of Erie Mining Company to collect maple sap, and he even described the method of tree tapping they employed, which is much different from how trees are tapped with metallic spiles today. In the interviews for chapter 5, John Koepke's work on the Laurentian Vision was surely informed by his Native heritage, and Koepke recalled an episode that occurred in deep winter, when he was encouraged by the Minnesota Department of Natural Resources to present his concepts to tribal departments of natural resources at the Black Bear Casino in Carlton, Minnesota. He said their initial reception was somewhat guarded until those groups came to better understand that the goal of the Laurentian Vision was to "put the ecology of the landscape first"—an ideology with which they felt a commonality and could potentially support. In November 2022 Darren Vogt of the 1854 Treaty Authority, an intertribal natural resource management organization, presented a synopsis of treaty rights to the current participants in the Mineland Vision Partnership (née Laurentian Vision Partnership). These interactions, though just three examples of a history that is largely unexplored in this book and that is worthy of its own examination, are evidence that the Indigenous people who reside on the lands of the 1854 and 1855 Treaties today continue to assert their interest in and vision for the use and future of the mined lands of the Mesabi.

Contents

Author's Note

In some ways, this book was written backward. As you will see, my personal and professional involvement with the mined landscape of the Mesabi doesn't begin until the period recounted in the last chapter, when, in the service of establishing the Redhead Mountain Bike Park, I had to learn to work within the complex mining, social, and political landscape of the Mesabi Range.

When Redhead was completed and began to attract attention from the public and the press, I was asked to write the story of the park's creation. In contemplating this undertaking, I began to realize the story of Redhead Mountain Bike Park was so inextricably linked to the earlier history of mineland creation and reclamation that its full telling would require a multichapter history. Thus, the book was born.

Minescapes attempts to trace the origins of many of the mining practices and laws that eventually had an effect on the park's creation, struggles, and ultimate legalization, funding, and construction. It is not intended to be a complete history of mineland reclamation on the Mesabi. Rather, it is a series of historical vignettes highlighting topics that remain important today on the landscape of one of the world's largest deposits of iron ore.

My principal technique in telling these vignettes was—to the extent possible—to focus on and interview a handful of people who were involved in state-of-the-art mineland conservation, reclamation, and repurposing for their respective time. This handful of people become, in a way, the characters of the story. Given that mineland reclamation did not develop in earnest until the 1970s, many of them are still around. For those who are no longer with us, I interviewed coworkers and family who knew them and extracted quotes from their published writings, letters, and other historical documents. This method intentionally focuses on the human side of an otherwise technical topic. By lifting up a small number of characters whose stories the reader can easily follow, I by necessity have omitted the names and stories of many, many others who had noteworthy contributions to the history of mineland reclamation and repurposing on the Mesabi Range. For that, I am sorry. I made this choice in the service of telling the story in an easily digestible way for a broad swath of potential readers who are interested in knowing more about the Mesabi lands after mining.

My second technique was to intersperse the historical narratives with vignettes of my own present-day visits to the landscapes in which these stories are set. The goal of these is to vivify the past and, in some instances, to track the progress of reclamation through the decades that followed its original execution and documentation. The write-ups of my visits are light on technical content—again, to make the story more easily read as a narrative history—and use largely open-minded, observational methods akin to being an amateur "naturalist of the unnatural." They should be considered nothing more than anecdotal observations of particular spaces in time from which broad-based scientific conclusions should not be drawn without further study.

Thank you for your interest in the story of reclaiming and repurposing one of the world's largest mined landscapes. It's a story that remains in its early stages, is still unfolding today, and will not be complete for at least another lifetime or two.

CHAPTER 1

Introduction

It's not an untouched wilderness like a mountain top, but a ramshackle wildness in which people and the land have conspired to strangeness.

Helen Macdonald, *H Is for Hawk*

It's 34 degrees Fahrenheit with occasional snow pellets, hard and round like pearl sugar, peppering the sidewalk in front of our two-story colonial revival home, yet the doorbell rings and rings. There is Cruella de Vil, an ax murderer, hockey players, princesses, hunters, a man loudly dragging a shovel and wearing a faux blood–spattered Tyvek suit with "CDC" handwritten on the back, and even a young Bob Dylan. They all rotate through our front porch. The sidewalk is clogged with parents, some in costume, some in hoodies, cigarettes in hand. The street itself is pandemonium, with trucks and cars weaving in and out from the curb like Manhattan traffic. There's even a pony pulling a two-wheeled tilbury festooned with multicolored LED lights. By 7:00 pm we have emptied our four Costco-sized candy bags and need to extinguish the front lights. The festivities will continue for several more hours and well into the darkness, even after most towns have quieted down. Our neighborhood's Halloween turnout astonishes year after year, and never more so than during a pandemic. All in all, nearly a third of the children in Hibbing, our small city of 16,077 residents—well over a thousand children—will visit our home for Halloween. On this night, I would compare the human throng on my small-town street to any boulevard anywhere, from Chicago's Miracle Mile to the Champs-Élysées. In our first year, my new neighbor Tom tried to prepare me, asking, "How much candy did you buy?" To my "a couple of bags," he pointedly replied, "You need *lots* more." Seeing the incredulity on my face, he simply said, "Welcome to Pill Hill."

Halloween scenes on Pill Hill, Hibbing. *Author's collection*

Pill Hill is a dump. To be more exact, it is a mine dump: a human-made mountain of earth and rock that was moved here at great expense from three miles further north. To the miners who hauled all 20 million tons of it here, it was just overburden, valueless material that needed to be moved to expose the rich natural iron ore deposit that later became the Susquehanna Mine. To the US Environmental Protection Agency, it would be "mine-scarred land," eligible to receive environmental cleanup funds. To the locals, it has been known by a series of names—factual names, aspirational names, and nicknames—that changed over time to reflect its shifting position and purpose in the minds of the people using it.

It was originally the Winston and Dear Dump. The prominent and unnatural appearance of the Winston and Dear Dump serves as the backdrop in many photos from the booming days of Hibbing. It appears as a steep-sided plateau behind a photo of Hibbing's auto raceway that

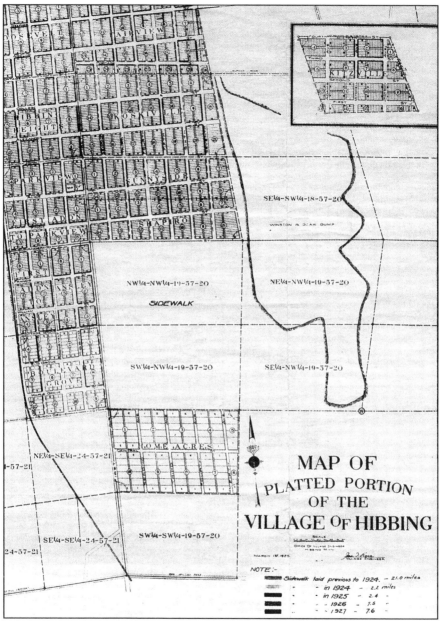

Map showing the Winston and Dear Dump adjacent to the platted portion of the Village of Hibbing, March 1, 1925. *Minnesota Discovery Center*

Hibbing High School with Winston and Dear Dump in background, mid-1920s to mid-1930s. *Aubin Photography Studios, Hibbing Historical Society*

"Air-dump" cars unloading on an elevated trestle, 1936. *Minnesota Historical Society collections*

shows goggle-eyed automobile racers spinning out on the dirt track like a frame from the children's book *Go, Dog, Go!*. You can see its variegated, treeless surface fingering out like a scallop shell or a walleye fin behind the town's famous high school, the $4 million "Castle in the Sky" that was completed in 1920 to entice residents southward and off the valuable iron formation. And you can see it being built in granular detail in a 1936 photo as railroad "air-dump cars" are emptied from an elevated trestle that would have been moved laterally in a fanlike pattern when deposited rocks reached the trestle's height, eventually creating the dump's scallop-like surface.[1]

In those times, Hibbing was known as the "Richest Little Village in the World" and was raising and spending more money on roads, schools, public works, and amenities than the state of Delaware. This same town was also reported as having "what is perhaps the ugliest townsite in the world" on account of "the huge mounds of debris [that] surround the community, wherever the eye may rest." The Winston and Dear Dump was once such a mound.[2]

But to the youth of the area, such as John Dougherty, Butters and Moose Kalibabky, Eddie Strick, Red Gilbert, and Geno Nicolleli, the

Ski jump built by residents on "the Dumps," Hibbing, 1938. *Minnesota Historical Society collections*

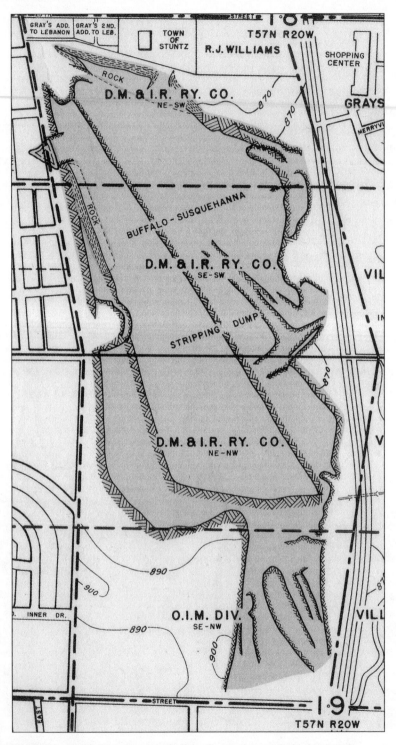

Detail from 1959 map showing Buffalo-Susquehanna Stripping Dump, Hibbing. *Minnesota Discovery Center*

Winston and Dear Dump was just known as the Dumps. According to Dougherty, the Dumps was an unregulated playground built upon "two tiers of overburden, stacked one atop the other, form[ing] this geological phenomenon." Using material stolen from construction sites, they built, as young daredevils of the 1930s and 1940s commonly did, a series of ski jumps, the largest of which could launch skiers 90 to 150 feet onto the landing below. The Dumps, to Dougherty, was no more than a ski hill upon which to strengthen the leg muscles that would eventually carry him down the Alps on the vacation of a lifetime.[3]

By 1959, under the ownership of the Duluth, Missabe & Iron Range Railroad, it was known as the Buffalo-Susquehanna Stripping Dump. Now it contained not only overburden but small pockets of lean ore rock along its the northern and western boundaries. The lean ore consisted of iron-bearing rock that was—at the time it was mined—too low in iron concentration to be considered "ore." The rusty-red hematite lean ore created a rockier, coarser surface to the northern and western edges of the dump than the mixed sand, silt, and glacially rounded boulders that were deposited elsewhere. Meanwhile, the city of Hibbing, which had been moved en masse southward to accommodate the expansion of mining on the original townsite, grew on all sides of the Dumps.

By 1960, the nature of the Dumps changed. The demand for new housing led to its residential development as Highland Park—but to Hibbing residents, it became known as Pill Hill in reference to the large population of doctors who settled there. That year it boasted a population of 128 residents. In contrast to its industrial, blue-collar birth, Highland Park now suffered no heavy industrial traffic, no railroads, no dirt roads, no poor streets, and no other "serious adverse or blighting influences." In this era, Pill Hill earned its reputation as a trick-or-treat destination worth traveling to. In the 1980s, it was a place where the parents of Boston Celtics star basketball player Kevin McHale invited kids into their atrium to select a full-sized candy bar from a rack that was outfitted like a convenience store.[4]

Today, Pill Hill remains much as it was decades ago. The houses erected in the 1960s still stand. The neighborhood still has just one access road; a second access road and additional platted residential lots lie, undeveloped, under a canopy of first-generation forest. The overburden is covered with aspen, a mowed-grass park, a water tower, and a volunteer-maintained trail weaving through wild apple, crab apple,

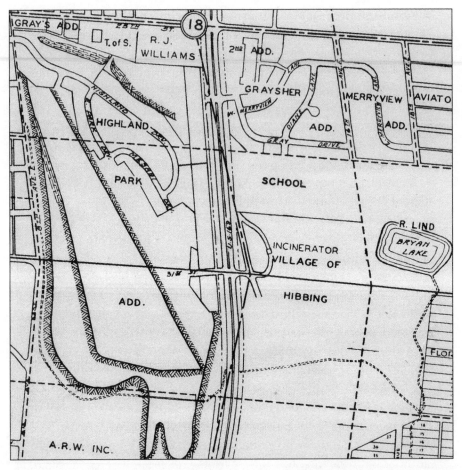

1960 map showing the residential
development Highland Park, Hibbing.
Minnesota Discovery Center

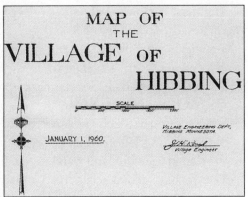

MAP OF
THE
VILLAGE OF
HIBBING

SCALE

JANUARY 1, 1960.

VILLAGE ENGINEERING DEPT.,
HIBBING, MINNESOTA.

Village Engineer

plum, and pin cherry trees. The vegetation covering the lean ore is different—pines, birches, and wild roses rather than aspens or fruit trees—and the rusty-red surface will still stain your dog's paws a permanent sanguine. Yet, on the economically stifled Iron Range, Pill Hill is considered a wealthy neighborhood. The yards are lush in turf and rich in seventy-foot maples. But when one digs a hole for a new tree or garden to be planted, one finds the unnatural origin of the development inches below the surface in the hard iron ore fragments and layers of upturned earth. At barbecues, neighbors recall Pill Hill's early mining origins and wonder whose house overlies the buried locomotive that fell irretrievably off the tracks into the muskeg beneath the overburden.

The story of Pill Hill, with its industrial beginnings before 1920, temporary recreational use in the 1930s and 1940s, and residential redevelopment in the 1960s, is unusual on the Iron Range. It is only one among 1,400 mine stockpiles on the Mesabi Iron Range that have been identified by the Minnesota Department of Natural Resources. Overall, there are nearly 130,000 acres of mine-disturbed lands on the range. The acreage that has been dug up, processed for iron, or deposited elsewhere is roughly equivalent to nine Manhattan Islands or the Twin Cities of Minneapolis and St. Paul plus the suburbs of Bloomington and Richfield. As of 2015, only 4,663 acres of this land, less than 4 percent, had been officially reclaimed; the remainder has been left alone or was still undergoing reclamation. A 1978 mining land use study reported that "in most

Pin cherry tree and wild roses growing on the lean ore of a former mining dump now known as Pill Hill neighborhood, Hibbing. *Author's collection*

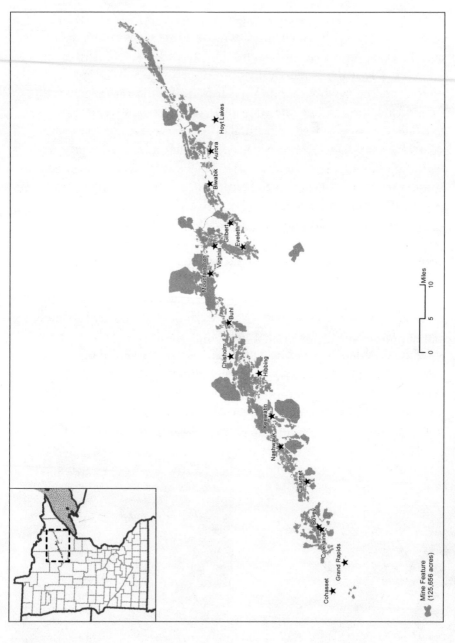

Mine-disturbed lands on the Mesabi Iron Range. *Courtesy of Jim Lind*

cases, exhausted mines, abandoned mining facilities, and surrounding land have not been reclaimed to any recreational or economical value." A few years later, mineland reclamation rules were enacted for new mining operations, but the lands that had been disturbed during the first eight decades of mining were exempt and thus excluded from purposeful restoration like what occurred on Pill Hill.[5]

What happens to the mining landscape of the Iron Range—both that which is purposefully reclaimed and that which is left alone—is the subject of this book. The qualities of these lands are often seemingly contradictory and difficult to reconcile. Unnatural, yet alive, like the annual Halloween celebration on Pill Hill. Disturbed, yet lush in certain species that are rare and highly prized elsewhere. These "double-truths," which defy many of the common narratives about mines and mining communities, are emblematic of the misunderstood history of Minnesota's mined lands. For example, tailings basins, often described as "moonscapes," have been documented as refuges for waterfowl as well as rare bird species dislocated by farming of their native prairie lands. Abandoned water-filled mine pits that in other parts of the country result in fish kills here on the Iron Range host both cold-water trout fisheries and award-winning drinking water supplies. Unreclaimed mine pits and piles have been transformed into world-class mountain biking venues, yet surveys of park users rank mining as one of the top negative influences on the Iron Range's future.

Like the mined lands themselves, people's opinions about the Iron Range landscape are diverse and seemingly contradictory. "Desolate and wasted" was how we described *our own* lands in the 1968 publication *Diamond Jubilee Days*. I've heard "interesting" and "beautiful" used by out-of-town visitors. Once a German tourist told me, as we were overlooking the "Grand Canyon of the North" (Hibbing's Hull-Rust open mine pit, which is on the National Register of Historic Places), "Trees are everywhere, but *this*, this can only be here." One lease manager from the Minnesota Department of Natural Resources said it "looked like money," referring to the mines' role in creating our nation and the fortunes of many millionaires. Sometimes the contradictions appear in the same piece of work, as in a strange black-and-white advertisement photo of a denuded open pit with locomotives puffing plumes of smoke accompanied by the phrases "the Pleasure Center of the Arrowhead

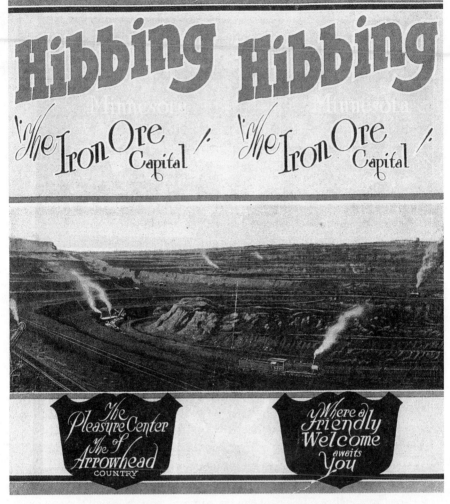

Industry and tourism: Hibbing represented as both the "Iron Ore Capital" and the "Pleasure Center of the Arrowhead." *Minnesota Discovery Center*

Country" and "Where friendly Welcome awaits you." Whatever your opinion of mining, it is hard to argue with the statement made in 1951 by exploration and mining pioneer Edmund Longyear that "few areas in the United States have been so completely altered by man."[6]

With Minnesota's first nonferrous metal mining permits issued in 2018, the state is entering a new era for the industry. The rancorous public debate about these permits makes clear that opinions on mining vary widely, from staunch opposition to stalwart support. It could be argued that one's opinion of the 130-year-old legacy of Minnesota's historically mined lands weighs heavily in the decision to support or oppose future endeavors. If the United States is to make the transition to a carbon-neutral economy that is not dependent on a foreign supply chain of critical minerals that could be disrupted by political unrest or a global pandemic, it is clear that more domestic mining will be required in the future. And more so than ever before, investors and mining industry leaders care about the social license for mining companies to operate. With the convergence of these factors, now seems like a good time to reevaluate the history of our mined lands and the reclamation efforts they have undergone. The stories and motivations of the people who created this complex landscape on the Iron Range help explain the seeming contradictions previously enumerated. From the earlier miners who, against the technological odds, developed the Mesabi Range but paid little heed to the postmining landscape to the modern-day proponents of mineland repurposing, the condition of the mined lands themselves records the value system of each generation that created, touched, and lived among them. This work explores the record that is written on the lands, from the boomtown days of late nineteenth- and early twentieth-century mining through regulation in 1980 to the modern visions of landscape architects, planners, engineers, and public officials.[7]

The intent of this book is to provide stories from various historical points that highlight advancements in the field of mineland reclamation and repurposing, focusing on the Mesabi Iron Range. The stories are intended to illustrate the thinking, the challenges, and the decision-making framework in place at the time in which each is set. The stories intentionally focus on the state of the art—the leading edge—of mineland reclamation and repurposing practices. Another set of stories could be focused on the lagging aspects of this practice, and in fact, many authors have dedicated their efforts to telling those tales. But the stories of deliberate actions taken by the mining industry, state and local governments, universities, and citizens of the Iron Range to conserve, study, reclaim, and repurpose mined lands are not well known, and they are worthy of being documented and examined in this era of public interest and debate over mining and the environment. They would be inspirational because they highlight a potential future for mined lands and a mining region that does not often receive attention. At the same time, the stories do not shy away from the challenges, and often the unintended consequences, that occur when that future is forged. Each vignette is intended to provide a wider-angled snapshot of mined lands than is typically offered—featuring not only the economic boon they fueled and the environmental disturbance caused by their exploitation but also the forms of life that can and do come back to them when mining is ceased and how that life struggles onward to give these lands a future. The fuller picture may help us understand and even appreciate the value and potential of the disturbed landscape of the Mesabi. By understanding it, we may grow to care for it in a new way. After all, as James Andrew Merrill, the geographer of the Lake Superior basin whose name is attached to a road near the Iron Range and the lake beside my cabin, wrote, "he who knows this region best, loves it most."[8]

CHAPTER 2

The Boom

It is also vandalism wantonly to destroy or to permit the
destruction of what is beautiful in nature, whether it be a cliff, a
forest, or a species of mammal or bird. Here in the United States
we turn our rivers and streams into sewers and dumping-
grounds, we pollute the air, we destroy forests, and exterminate
fishes, birds and mammals.

Theodore Roosevelt, *Outlook Magazine*, January 25, 1913

On the sun-beaten plains of North Dakota that stretch on endlessly
beneath my wheels, I try to imagine making this trip on a horse. How
would my hip sockets feel after hundreds of miles of prairies and in-
numerable river crossings? Or how would my back muscles knot up
from rowing the Missouri River in a collapsible boat like Lewis and
Clark, whose trail we have joined? In my air-conditioned compart-
ment, it seems hard to imagine how our forebears did it, just a few
lifetimes ago.

As for my family and me, our vehicle is an Odyssey. It's a durable
steel, plastic, and rubber wonder—a quarter of a million miles on the
odometer, yet somehow still smoothly assailing I-94. Neither our hip
bones nor our back muscles cry out, yet our bodies still require a rest
stop. It is the first week of summer vacation for the children. Our crew
of seven has left the land of mines and pines for the seas of soy and corn.
This flat, fertile soil is foreign to all of us, having been born in Negaunee,
Michigan; Hibbing, St. Paul, and Waseca, Minnesota; and Lusaka, Zam-
bia. We are heading west, the direction of adventure, to Big Sky, Montana, where I will deliver a speech to the American Society of Mining
and Reclamation and then we will enjoy a family vacation.

We blinker our exit, and just where the flatland cracks open to glo-
rious, striated pink, buff, and tan canyons, we see him. Statuesque. A

brown bristle of hair, a powerful chest—majestic, yet wild and western. A quiet, intelligent eye belying the bully he would become when driven. This is Theodore Roosevelt—the national park, that is—and the bull bison who casually greets us at the gate.

Beyond the Badlands, Roosevelt shows up again on Mount Rushmore. We soon see photos of Roosevelt camping in Yellowstone. There is Roosevelt, Utah; Roosevelt, Arizona; and Roosevelt, Washington. Tributes to the former president scatter the land like Paul Bunyan's giant footprints; people love the man who so loved the western land and fought to conserve it—the Wilderness Warrior, Theodore Roosevelt.

■ ■ ■

It is hard to imagine that Roosevelt, our celebrated conservationist president, had anything to do with the broken landscapes of the Minnesota Iron Range. How could the man who, in 1913, lamented the creation of dumping grounds possibly be attached to a land whose principal feature is mine dumps?

Although he frequented St. Paul and Minneapolis, visited Winona, and even hunted birds in northwestern Minnesota, there is no evidence that Roosevelt ever visited the Iron Range. During the peak of his popularity he was, of course, *invited* to visit the Iron Range to see the mines and to hunt moose. But a document search shows no evidence that Roosevelt, a prolific writer, ever even wrote the name Mesabi—a surprise given the fact that the range rose to prominence as the nation's single largest source of iron ore during his period of political power and at a time when steel was essential to undergirding the infrastructure of the growing nation. Had he ever visited the Mesabi Range, I believe it would be as well known as his famous sojourns to North Dakota and Montana.

What Theodore Roosevelt did have on the Iron Range was a friend, and friendship can be a powerful thing. For inside friendships, like that of Thomas Edison and Henry Ford, discussions occur in which ideas take shape. An encouraging word from a trusted friend can cement those ideas into actions. And those actions can leave a lasting imprint on the land, such as by the declaration of a national forest or the digging of an open-pit mine.

It's through the story of friendship that I'll examine the early days of the Iron Range in the first decade of the 1900s. These were the days when steam shovels and ore beneficiation plants changed mining from a small-scale artisanal pursuit to a landscape-altering industry. They were also the days when conservation became a national movement and over 150 million acres of national park and forest lands were established. Sometimes, these consequential land actions took place within just a few miles of one another. Often, they resulted in bitter conflict. But only here in Minnesota's Arrowhead region were they simultaneously led by two men who considered themselves to be lifelong friends. The story of this friendship—and what would seem to be a surprising unity of opinion and outlook in men whose actions had such vastly different outcomes for the landscape of northeastern Minnesota—begins in the sultry air of Cuba.

■ ■ ■

Under the tropical sun, two men lie sweating in their uniforms. The temperature is nearing 100 degrees Fahrenheit and the jungle humidity is stifling. Both men are American officers—one aged twenty-six, the other almost forty. After months of training, and now several days into a largely unopposed invasion, they wait for action, for "the crowded hour when the wolf rises in the heart." But they are pinned to the bottom of a hill, held in place by enemy gunfire and positioned behind their comrades. Not a good spot for glory.

Their moment comes when the .30-caliber American Gatling guns open up, discharging seven hundred rounds per minute into the Spanish battle lines, and the men charge the hill. Though they are cavalry soldiers, they run on foot, passing their crouching colleagues. The younger man—a second lieutenant and a two-hundred-pound, six-foot-tall football and baseball star from Yale—leads the burst. He is meant for this type of athletic display, the one man selected from every twenty applicants for this position when this particular call to arms was posted. His youth and strength propel him to the top of the hill, where he is the first American officer to leap into the enemy entrenchments. The other man, his colonel—chestier, older, and already having led the charge up one hill (Kettle Hill) but no stranger to bold and strenuous physical feats—remarks with envy, "I wanted to be the first there myself, but he outran me!"

The date is July 1, 1898. The regiment is the Rough Riders. And the hill is San Juan. After the battle, the younger man will be awarded the

Theodore Roosevelt (center) and the Rough Riders, San Juan, Cuba, 1898.
Houghton Library, Harvard University

Silver Star and promoted to first lieutenant for his act of gallantry. The older man, having led his troop of volunteer cavalry soldiers to victory, will become "the most acclaimed man in America." The men, whose lifelong friendship was cemented on that battlefield and others, are John Campbell Greenway and Theodore Roosevelt.

Greenway and Roosevelt had come to Cuba from much different places. Greenway, still a young man, had left a position with Carnegie Steel Company (later U.S. Steel) in Duquesne, Pennsylvania. Despite holding an engineering degree, he had begun his career "at the bottom," earning $1.32 a day by helping feed the world's largest steel blast furnace, "Dorothy Six." His celebrity stemmed from his athletic accomplishments as an end for the national-champion Yale football team and as a catcher on the baseball team.

Roosevelt, for his part, had already lived through his early career and begun climbing the political ladder from member of the US Civil

Service Commission to New York City police commissioner to assistant secretary of the navy. By 1898, he was a rising political star at midcareer who, against the better judgment of his friends and political peers, had taken a break from office seeking to help his country wage war against Spain. In the first half of his political life, Roosevelt had already sown the seeds of a conservationist movement in American politics—cofounding, among others, the Boone and Crockett Club of preservation-minded outdoorsmen with George Bird Grinnell (who would later play an important role in establishing Glacier National Park) and taking train rides across the country with the likes of naturalist John Burroughs, heralding the beauty of the unspoiled parts of the West and waging political war against the logging interests and mining companies that threatened to encroach on the boundaries of Yellowstone National Park.

Greenway and Roosevelt's mutual love of country and adventure brought them together in 1898. After the sinking of the USS *Maine*, both men joined the war; in Texas, they trained together as the Rough Riders grew into a fighting force made of the strange admixture of college athletes and Western cowboys. In camp, they discovered they both were men of science, Roosevelt specializing in biology and Greenway in engineering. Both were progressive Republicans. Both loved athletics, Greenway opting for ball sports and crew while Roosevelt reveled in boxing and wrestling. Both were Ivy Leaguers. They shared moral ideals about how men should lead by example, treat their subordinates, and contribute to the growth of their country. As the commanding officer, Roosevelt felt he could trust Greenway with difficult tasks, once ordering him to fight and dig trenches without rest for forty-eight hours straight.

After the battle of San Juan Hill and subsequent routing of the Spanish from the Cuban capital of Santiago, both Roosevelt and Greenway remained fit and healthy during the weeks of occupation (while half of the other Rough Riders were incapacitated by wounds, malaria, and yellow fever). During this idle period, Roosevelt convinced Greenway to venture a three-hundred-yard swim in the Caribbean Sea to visit the wreck of the USS *Merrimac*, which lay on the bottom of the harbor, having been disabled by Spanish guns. As they departed the shore, the men began to notice the waters around them darkening with large, piscine bodies. Greenway grew alarmed when he realized they had attracted a school of sharks. In a moment that illuminated his fearless nature, Colonel Roosevelt swam on, reassuring Greenway between strokes that "*They—won't—bite!* I've been—studying them—all my life. And I never—heard of one—bothering a swimmer. It's all *poppycock*!" Eventually Roosevelt and Greenway reached the wreck. After completing the shark-circled swim, Greenway reported he felt better—until he remembered that he had to swim back.[1]

Perhaps not unexpectedly for men who swam with sharks, their trust and fondness for each other grew rapidly into what would become a lifelong friendship and correspondence. Their future public comments and writings would be full of praise for each other; Roosevelt mentioned Greenway three times in his autobiography, and Greenway celebrated Roosevelt by naming streets after him in the towns that he founded. And as they sailed away from Cuba in August 1898 toward their debriefing and recuperation in Montauk, New York, in similar ways the war in Cuba set them both on the path of being public figures, leaders of men and movements, and important decision-makers. The principal difference was scale: Roosevelt operated on a national level and Greenway on a regional one.

■ ■ ■

In that same summer of 1898, another ship set sail halfway across the world. This one carried not war heroes but emigrants fleeing poverty and the threat of a dictatorial edict that would conscript young men into the army of the czar they cared nothing about. The emigrants, mostly men leaving their wives and young children at home in the Russian Grand Duchy of Finland, were heading to America and the iron mines there that promised work. In their home country, they were landless tenant farmers who adopted the surname of whatever farm they happened to be working. They had no mining experience and spoke no English, but they had heard that in the familiar-looking northern lake country of the United States land was available to men who would work in the dangerous underground mines. Unbeknownst to the passengers, two of them would soon be among the three thousand miners under the assistant superintendence of John C. Greenway in the underground iron mining district of Michigan. Those men were my great-grandfather Matti and his brother Aleksanteri Kero.

In the years that followed, both Roosevelt and Greenway rose in

Matti Kero, the author's great-grandfather, reunited with family in Negaunee, Michigan, circa 1903. *Author's collection*

Aleksanteri Kero, the author's great-uncle, in Lapua, Finland, shortly before his emigration, circa 1898. *Author's collection*

station. Roosevelt's national acclaim from his service in Cuba propelled him to the governorship of New York and eventually the vice presidency of the United States. When President William McKinley was assassinated in 1901, Roosevelt was thrust into the Oval Office. He had gone from being one of the oldest volunteer soldiers of the war to the youngest man ever to occupy the White House.

Greenway's rise was less spectacular, but he and Roosevelt kept in regular contact. By January 1902, Greenway had shifted from blast-furnace helper to, in his own words, "a struggling employee of the U.S. Steel Company as assistant to the Superintendent." He invited Roosevelt to Upper Michigan to "let me return the compliment of the walk you gave me last fall along Rock Creek by giving you a climb through 'the stopes' of some of our deep mines a thousand feet underground." However alluring a thousand-foot underground climb may have sounded to an advocate of the strenuous life, Roosevelt never made the trip. Nonetheless, Greenway assured him of his politic support, stating, "It gives me the greatest pleasure to be able to help you carry out your ideas and purposes even if only in a small way."[2]

Three years later, at the beginning of 1905, Roosevelt won election to his second presidential term. In his first term, he had developed a reputation for bold, unprecedented conservation of American wildlands such as Pelican Island off the Florida coast. He had established thirteen new national forests. As he began his second term, he had designs to expand his conservationist agenda. He was soon to become what modern historians such as Douglas Brinkley would call a "conservation visionary, aware of the pitfalls of hyper-industrialization, fearful that speed-logging, blast-rock mining, overgrazing, reckless hunting, oil drilling, population growth, and all types of pollution would leave the planet in biological peril."[3]

For his part, by the turn of 1905, Greenway had become, using Brinkley's terms, a captain of "blast-rock mining." He was now secure in his position as assistant superintendent of the Marquette Range of iron mines for the Oliver Iron Mining Company, a subsidiary of U.S. Steel. Under his assistant superintendence were dozens of underground mines excavating 60 to 70 percent pure iron ore for shipment to the "insatiable" iron furnaces in the Great Lakes region. Demand for steel was growing, for the "rails that span the continent," "the girders that rear aloft the skyscrapers," "the pots and pans in the cupboard," and "the keys in one's pocket"—to quote Duluthian J. S. Pardee in his unpublished manuscript "Of the Womb of Wealth." Greenway's position put him in the center of the Lake Superior district, which, as Pardee describes, is a great ellipse of iron ore deposits "forming a necklace around Lake Superior," with merchantable ores being mined in Upper Michigan, northern Wisconsin, and northeast Minnesota. At this time of enormous steel demand, a full four-fifths of the nation's iron was mined in the district.[4]

As a rising star of industry, Greenway was active in the community and was well respected by men who worked for him—like my great-grandfather. (I sometimes wonder whether Greenway was one of the "dignitaries of Negaunee and Ishpeming" who my uncle said filled my great-grandfather's sauna on Saturday nights.) He had also distinguished himself as an engineer and inventor of the Greenway Ore Unloader, a mechanism to prevent the snows of Upper Michigan from freezing the iron ore into the railcars, for which he received

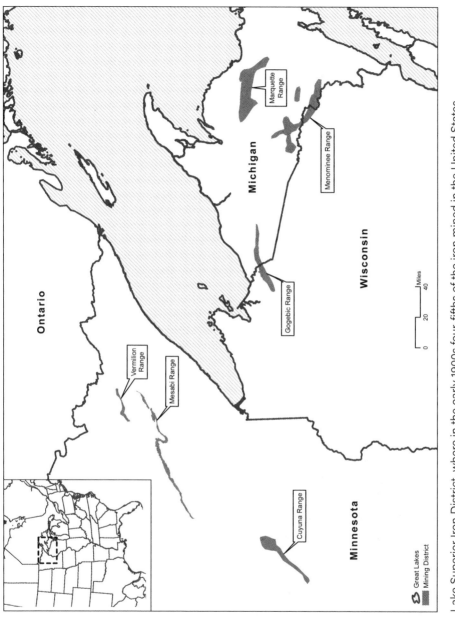

Lake Superior Iron District, where in the early 1900s four-fifths of the iron mined in the United States originated. *Courtesy of Jim Lind, modeled after a figure by John Baeten*

both a patent and a personal note from Roosevelt, who was pleased and "de-lighted" (one of his famous catchphrases) with the invention. Greenway's community-mindedness and ingenuity were to play an important part in his next assignment.

Greenway was about to be sent to Minnesota and the western end of the great Mesabi Iron Range, the world's largest known deposit of iron ore and the seat of three-fifths of domestic iron mining, to oversee a new venture for Oliver. It was to be the largest digging operation conducted for iron mining at that time. But more importantly, his charter was to develop a technique for the purification of Mesabi "natural iron ore" that had not yet been tried in the Lake Superior iron district.

At the time Greenway was reassigned to the Mesabi Range, it had already gone through two phases of iron mining: underground mining (similar to the Michigan iron ore mining that Greenway administered) and small-scale, open-pit mining (sometimes referred to today as "artisanal mining"). Both types were profitable because of the high purity of the Mesabi iron ore—it contained a concentration of more than 57 percent iron and was called "direct ship" ore because it simply required exposing it, digging it up from the ground, and shipping it to the furnaces and steel mills of the lower Great Lakes. Not only was it profitable, but by 1905 it was such a bounty that Chester Congdon, lawyer for U.S. Steel and landowner on the Mesabi Range, had already laid the foundation for his mansion in Duluth, Glensheen, which would cost $854,000 (the equivalent of more than $21 million today).[5]

By this time, it was clear that the Mesabi Range was special among mining districts. Although iron is the fourth-most abundant chemical in the Earth's crust—adding its hallmark red tinge to such well-known geologic formations as Utah's Grand Staircase–Escalante and Colorado's Red Rocks amphitheater—its usefulness as an ore depends on the ease with which it can be extracted and purified. During this period, the predominant steelmaking operations in the United States (which used the Bessemer process) required ore to be over 50 percent pure iron to be valuable; in comparison, the red sandstones of the American Southwest are usually just a few percent iron. The banded iron formations of the Lake Superior region were, and still are, the most economical location in the country to find such high-purity ores. These rich ores were created in what geologists believe to be a once-in-a-planet occurrence, best described by science writer John McPhee: "The transition that produced them—from a reducing to an oxidizing atmosphere and the associated radical change in the chemistry of the oceans—would be unique. It would never repeat itself. The earth would not go through that experience twice."[6]

Although this process created iron ore deposits all over the world, it was the sequence of two geological transformations that turned the iron-bearing Mesabi host rock into a world-class ore. The first occurred when water entered cracks in the host rock, removing impurities and concentrating the iron. J. S. Pardee poetically describes it this way: "Water dripping through the rock carried away silicon and brought more iron, till the rock itself in some places was changed into ore, molecule by molecule." This natural leaching process created direct ship ore. Pardee goes on to say that "in places the story of the formation can be read in the beds of ore. The source of an underground stream is sometimes distinctly traced. Its pebbles are found buried in ore just where the water left off polishing them. In places the beach of [an] ancient sea is discovered, and teeth of sharks and other antediluvians have been found. Shells, once lime and silicon, have been made over into ore, preserving the mould but none of the substance of its original creature." While today's geologists may describe events differently, one can indeed find shark teeth in Hill Annex Mine State Park and stromatolite fossils on the modern-day mountain bike trails that crisscross the Iron Range.[7]

In the second transformation, glacial scour and other erosion events made the Mesabi ores very shallow, often sitting only a few feet below the land surface, which is why some explorers discovered iron ore under the upturned roots of wind-toppled trees. This shallowness was critical in making the iron ore easy to mine. In Pardee's bombastic words, "except that the glaciers had reamed off that rock, gentlemen would be shaving themselves with clamshells to this day and ladies sewing with kits of bone."[8]

As the ironmaking furnaces adjusted to the soft Mesabi direct ship ore, land speculation and mining in northeast Minnesota became a race. In the 1890s, when the euphoria of vast iron ore discoveries was dawning, the *Duluth Evening Herald* reported, "Sensation has followed sensation with such rapidity that the public mind has had no time to recover from one shock until it has had another and is now prepared for almost anything." Mines sprang up everywhere. "Location towns," which were small collections of homes and businesses created so miners could easily walk to the locations where they worked, swelled in population to become small cities. Legend had it that the mineral wealth created more millionaires per capita in Duluth than anywhere else in the nation based largely on the profitable Mesabi ores (a claim that has not been conclusively proven). The only problem with direct shipped ore was its limited supply. By 1908, Andrew Carnegie went on record as saying that domestic iron supplies would be "exhausted in 40 years, probably in 25." Whether this statement was made objectively or motivated by business interests involving foreign iron ore tariffs is subject to debate. Nonetheless, the impact was clear: steel men were beginning to understand that the world's most sensational iron ore deposit had limits.[9]

This is where John C. Greenway stepped in. A vast untapped reserve of iron ore lay in the western Mesabi, but it was mostly too sandy to be a good direct shipping ore. Although some seams of high-purity ore existed, the iron concentrations in the sandier ore zones ranged from 40 to 49 percent, which was not considered merchantable. Greenway's challenge was to convert this siliceous ore into a product usable at U.S. Steel's furnaces, thereby expanding the domain of the Mesabi Range to the west, developing new communities to support the mining operations, and extending the Oliver Iron Mining Company's ore reserves.[10]

When he moved to the Mesabi territory in spring 1905, his first job was to survey the extent of the ore under Oliver's control. It proved to be a large and thick, but relatively low-quality, deposit of banded iron ore that lay tilted like a pillow upon a low headboard of granite known as the Giants Range batholith. The ore was covered by a layer of sand and gravel that had been left in place by the glaciers that retreated from Minnesota at the end of the Ice Age. Geologists often refer to this layer as glacial drift—a catchall term for everything the mile-high glaciers churned up, carried along, and deposited during the slow course of their advances and retreats across Minnesota.

The Michigan mines Greenway administered were underground networks of interconnecting shafts, adits, and stopes created to access the deeply buried ore. By contrast, Oliver's western Mesabi deposits were relatively shallow and flat, dipping only slightly from north to south, which made them suitable for mining through the excavation of large, open pits. For the open-pit mining to proceed, however, the glacial drift needed to be removed to prepare the mine for ore extraction—a process known as stripping.

Greenway's survey revealed that stripping the western Mesabi would require the removal of over 10 million tons of glacial drift. This material the glaciers left behind would become waste, deposited in piles around the mine known as stripping dumps. Because the drift lies over the valuable ore and was costly and troublesome to remove, miners typically refer to this material as overburden. Hundreds of acres of overburden would need to be systematically dug, loaded, and redeposited into human-made mountains of drift. The upper material—topsoil, or sometimes swampy muskeg—was removed and placed at the dumping ground first, followed by the underlying layers of clay, sand, and gravel. It was as if the surface of the western Mesabi would be "flipped like a pancake" to expose the ore beneath. This flipping action left the good black dirt at the bottom of the pile and elevated the nonorganic sand, gravel, and clay to the top, making it difficult for vegetation to quickly reestablish itself on the stripping dumps. At the time, Greenway's was to be the largest stripping operation ever undertaken by Oliver.[11]

Stripping required a massive earthmoving operation that would not have been possible without the key technological change that was happening in mining around the turn of the century: the development and use of the steam shovel. This machine allowed surface overburden to be stripped from the iron mines at rates impossible for artisanal mining. The general philosophy of steam shovel–enabled mine stripping was to remove and dump the waste materials as quickly as possible. In comparison to construction earthwork, mining operations were, as a 1920 textbook described, "of a destructive character and much less attention need usually be paid to the aftermath of the excavations or to the final disposal of the debris. The direct object is the winning of the ore, efficiently and cheaply." The steam shovel removed and deposited overburden at speeds not seen before. To use the textbook's terminology, the rate of "destruction"—that is, removal of material and "final disposal" of wastes into stockpiles—occurred at 5 million tons per year. Far to the south, the steam shovel techniques that were to be employed on the western Mesabi were also being used to dig one of Roosevelt's passion projects, the Panama Canal.[12]

Greenway was also studying the means needed to purify the sandy

Theodore Roosevelt digging in the Panama Canal, 1906. *Theodore Roosevelt Center, Dickinson State University*

ore of the western Mesabi. Knowing that any beneficiation process would require a lot of water, he advised his boss and vice president of Oliver Iron Mining Company in 1905 to secure ownership of the shoreline of Trout Lake. According to historian Donald Boese, who documented the western Mesabi stripping operations in his *John C. Greenway and the Opening of the Western Mesabi*:

To William Olcott, Greenway wrote on February 11, "Yes, we should own all the land at the north end of Trout Lake." Greenway knew that any concentration project was going to require large supplies of water and that there would be vast quantities of waste products [because the ores there were too sandy for direct shipping, unlike the ores east of Nashwauk]. He explained to Olcott that it would be wise to own the lakeshore so when the inevitable contamination of the lake began, the company would not be faced with lawsuits from any property owners in the vicinity.

Thus, by the middle of 1905, Greenway was laying plans for what was to be an altogether new type of mining operation, with massive steam-powered stripping and water-based purification for the sandy ores. Each of these innovations resulted in unprecedented volumes of mine wastes—stripping dumps and washing tails—that would have a permanent effect on the topography and future landscape of the Iron Range.[13]

■ ■ ■

It was during this time, in fall 1905—when Roosevelt's conservation movement was empowered by his first victory as an elected president and Greenway's mining operation was poised to generate unprecedented levels of mine waste across the Mesabi Iron Range—that the two friends had the opportunity to reconnect in person. Roosevelt invited Greenway on a trip across the southern United States from mid-October to early November, consisting of the president touting his policies at stops in North Carolina, Georgia, Florida, Alabama, Arkansas, and Louisiana. Greenway would have been in the audience to hear the president roar the following words on October 19 to a crowd at Raleigh, North Carolina:

> each generation works not only for its own well-being, but for the well-being of the generations yet unborn, and if we permit the natural resources of this land to be destroyed so that we hand over to our children a heritage diminished in value we thereby prove our unfitness to stand in the forefront of civilized peoples. One of the greatest of these heritages is our forest wealth. It is the upper altitudes of the forested mountains that are the most valuable to the Nation as a whole, especially because of their effects upon the water

Theodore Roosevelt addressing a crowd in Raleigh, North Carolina, October 19, 1905. *Houghton Library, Harvard University*

supply. Neither State nor Nation can afford to turn these mountains over to the unrestrained greed of those who would exploit them at the expense of the future. We can not afford to wait longer before assuming control, in the interest of the public, of these forests; for if we do wait, the vested interests of the private parties in them may become so strongly intrenched that it may be a most serious as well as a most expensive task to oust them.[14]

Roosevelt's speech against the "vested interests of the private parties" and the "preservation of forests" may lead us to question whether Greenway had any misgivings with respect to his assignment on behalf of U.S. Steel in the pinelands of the Mesabi (Greenway's journal for this day recorded nothing at all). But just three days later, before a crowd in Atlanta, the president almost seemed to answer that question as he spoke of the "well-meaning, but misguided effort to check corporate activity" in Puerto Rico and the Philippines, stating that "there is nothing that the islands need more than to have their great natural resources developed." This statement would imply endorsement of the kind of resource extraction Greenway was preparing to conduct in Minnesota.[15]

Throughout their trip, Roosevelt spoke on the themes of natural resource conservation and the need to protect forests and water. During all of the speechmaking, the Roosevelt-Greenway friendship showed no evidence of flagging. In a manner that seems difficult to imagine amidst today's bitterly divided public opinions about mining, relations between the great conservationist president and a man later inducted into the National Mining Hall of Fame were harmonious. Evidence of their continued mutual admiration occurred in Greenway's home state of Arkansas, where the president took time to praise Greenway to the local newspaper. Reflecting on their bond, the *Arkansas Gazette* reported that "a close friendship exists between the two, and when President Roosevelt decided to come to Arkansas, one of the first things he did was to invite Greenway to accompany him. . . . 'We will have a bully time if you come along' was the message the president sent by a friend to young Greenway."[16]

After roughly three weeks together, the president returned to Washington, DC, while Greenway traveled to the southwestern United States to survey mining lands before returning to the Mesabi Range. Less than a year later, Roosevelt offered Greenway the commissionership of the General Land Office, the precursor to the Bureau of Land Management—a position from which he would have governed the leases, sales, and uses of federal lands, including mining and timber interests. In offering Greenway this job on November 6, 1906, Roosevelt wrote, "I want you to realize that this is one of the most important positions in the Government and I know of no position just at the time where a greater public service can be rendered. I need hardly say again how glad and proud I should be to have you connected with my administration." A fascinating alternative life may have emerged for Greenway had he accepted the president's offer, but he refused. Roosevelt settled for Richard Ballinger to fill the position but later wrote Greenway that "he can't be quite as good as the individual to whom I first offered it!"[17]

Greenway commenced stripping operations on the western Mesabi project, then known as the Canisteo district, in 1906. Such work was often handled by a contractor, but for this important project, Greenway oversaw operations himself. Stripping was conducted by rail with

Theodore Roosevelt (center), John C. Greenway (right), and a friend on a rail tour of the South, 1905. *Houghton Library, Harvard University*

steam shovels loading "dinkeys"—small locomotives—and dump cars on an ever-shifting network of roadbeds and track. The rails needed to be moved whenever the shovels hit the end of their reach and dumps were full to the trestle. The work was scrupulously planned and tracked with an eye toward finances, for glacial drift was waste and not paydirt.

In the words of Donald Boese, "The dumping grounds were carefully selected areas where there was no appreciable amount of underlying ore, but they also had to be located as close to the mines as possible because the cost per mile of track was above nine thousand dollars."[18]

Glacial drift was not the only material that needed to be moved during the mining of the Canisteo. Some of the rock layers contained iron at concentrations that, while much higher than typical bedrock found around the country, did not meet the minimum 50 percent purity that was needed for Bessemer steelmaking at the time. Some of these layers would prove useful to Greenway, whereas others contained too many fine particles or too low an iron concentration to be purified at the time. Classifying and disposing of this material, collectively known as lean ore, is an important hallmark of Greenway's approach.

Recognizing that these materials contained a high iron content that might someday make them usable, Greenway stockpiled the lean ore in hopes that better purification techniques would later be developed.

A steam shovel removing overburden in the Canisteo district, 1906. *Itasca County Historical Society*

The types of lean ore present in the Canisteo district included Cretaceous ore (named after the geological era when it was deposited), "paint rock," and taconite. Boese described the issue thus:

> For the mining men, the Cretaceous material posed a problem. Particles of iron had washed into the Cretaceous sea just as they had into the earlier sea of two billion years ago, and fairly high iron content was present. This material with its density and conglomerates could not be considered wash ore, nor was it high enough in iron concentration to be classified as a direct shipping ore. If, as seemed increasingly likely, open pit mining were decided for the district, the Cretaceous deposit could not be simply wasted and discarded along with the glacial drift. Rather, although involving additional expense, it would have to be stockpiled to await some future method of extracting the iron. . . . While high in iron content, the Paint Rock was of a dense consistency and had a greasy feel when wet. Like the Cretaceous layer, it was too rich in iron to be discarded, but it obviously could not be classified as wash ore. Although very difficult to handle, it too would have to be either stockpiled or bypassed by shafts.[19]

Boese's description bears repetition and emphasis. Mining operations such as Greenway's, which carefully planned and measured the expense of every mile of railroad track and had yet to turn a profit on the Canisteo, would bear the extra expense of segregating and saving the lean ore in dedicated stockpiles. This act of segregating and saving lean ore begins to hint at an important—and sometimes forgotten—tenet of conservation thinking at the time. Over the course of the two years it took for Greenway to strip the Canisteo, Roosevelt and his chief conservationist ally, Gifford Pinchot, would explain this line of thinking as they elucidated the principles of conservation in writings and speeches.

By 1908 Roosevelt began to directly define conservation as a philosophy. In an address to the only Conference of Governors held at the White House, May 13–15, 1908—which included James J. Hill and Andrew Carnegie, both major land and mineral rights holders on the Mesabi—Roosevelt gave a speech to the assembled leaders entitled "Conservation as a National Duty." It was here he uttered a now-famous quote: "We have become great because of the lavish use of our resources. But the time has come to inquire seriously what will happen when our forests are gone, when the coal, the iron, the oil, and the gas are exhausted, when the soils have still further impoverished and washed into the streams, polluting the rivers, denuding the fields and obstructing navigation." Concern over the exhaustion of natural resources was a central pillar of conservationism. Parsing the statement above suggests the president was concerned about the loss of the resource potential of the forests, soils, and minerals more so than the pollution of rivers. Pinchot expanded on these themes in an essay published in 1908: "It is true that some natural resources renew themselves while others do not. Our mineral resources once gone are gone forever. It may appear, therefore, at first thought that conservation does not apply to them since they can only be used once. But this is far from being the fact. Methods of coal mining, for instance, have been permitted in this country which take out on the average but half of the coal. Then in a short time the roof sinks in on the other half, which thereafter can never be mined."[20]

Roosevelt best summed up this principle—that resources could and should be developed in a deliberate but not wasteful manner—in his message to Congress delivered just six months earlier, on December 3, 1907: "I recognize the right and duty of this generation to develop and use the natural resources of our land; but I do not recognize the right to waste them, or to rob, by wasteful use, the generations that come after us." In this context, it is possible to begin squaring the notion of early conservationism with the massive landscape disturbances that Roosevelt's agreeable friend Greenway was directing on the western Mesabi. Roosevelt saw conservation in a literal sense—*conserving* our resources rather than using them lavishly. The emphasis was on the avoidance of wanton destruction rather than complete preservation from destruction, with the exception of certain historical landmarks, areas of scientific interest, and national parks that Roosevelt preserved through the Antiquities Act.[21]

Viewed through this lens, Greenway's creation of lean ore dumps was a conservationist action, in solid agreement with the thinking of the time—including that of his good friend. In the process of developing the iron ore resources of the western Mesabi to serve the expanding US economy, U.S. Steel incurred extra expenses of extracting, segregating, and preserving the ores that were not economical to process at the time (for use by future generations, who would presumably have better purification technologies). Contrary to Pinchot's example of wasteful coal mining that used only half the coal, Greenway's mine attempted to use as much iron ore as possible and preserved the unusable ores for the next generations. It was probably thought that, at the end of Carnegie's twenty-five or forty years, these final lean ore piles would be swept up and the Mesabi would be clear of stockpiled lean ore. It is doubtful Greenway could have dreamed that his Cretaceous and paint-rock stockpiles would remain largely untouched 110 years later as part of the panoply of mining landforms that created the modern landscape of the Mesabi.

The early conservationists relied heavily on the concept of "scientific management." As long as resources were used judiciously, with care taken to eliminate waste and preserve the option for enjoyment by future generations, development was encouraged. Pinchot later wrote, in *The Fight for Conservation*, "The first principle of conservation is development, the use of the natural resources now existing on this continent for the benefit of the people who live here now." Greenway's mine certainly achieved this objective, serving the exploding demand for steel in the United States at the turn of the century. Pinchot went on: "In the second place conservation stands for the prevention of waste." Greenway's lean ore stockpiles attempted to preserve the iron-containing rock units that had to be removed to access the profitable ore bodies of the Canisteo district. Only Pinchot's third principle, "that natural resources must be developed and preserved for the benefit of the many, and not merely for the profit of a few," might be questioned. By 1908, nearly all Mesabi minelands and mineral rights had been consolidated in the hands of a few individuals and corporations, with the possible exception of school trust lands.[22]

While Roosevelt and Pinchot were fighting to establish conservation across America's natural resource industries, there were, of course, many who believed that any thought of the future welfare was a waste of money. This attitude is best summed up in the words of western copper magnate W. A. Clark, whose mineral-derived wealth became infamous in the 2013 book *Empty Mansions*. Clark stated that "in rearing the great structure of empire on the western hemisphere, we are obliged to avail ourselves of all the resources at our command. The requirements of the great utilitarian age demand it. Those who succeed us can well take care of themselves." But Greenway seems not

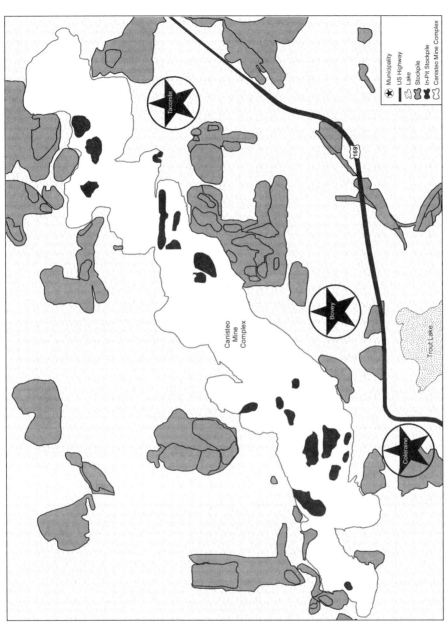

Stockpiles left after mining of the Canisteo mine complex. *Courtesy of Jim Lind*

to have been swayed by this line of thinking from some of his mining colleagues. In correspondence, he reassured Roosevelt of the rightness of the latter's conservation actions. After informing his friend that "my job here is moving along well, and I have a fine young organization of hustlers than can't be beat," he went on to write that "the country seems to be prospering, and the corporations earning plenty of money in spite of your 'attacks' on them."[23]

By late 1908, after two years of steady work, Greenway completed the stripping and lean ore stockpiling from the Canisteo district. The paydirt was exposed and ready. By the end of the operation, Boese writes, "The impressive total of 9,126,158 cubic yards of overburden [was] removed . . . at an expenditure of almost $2.5 million." For its $2.5 million investment, the Oliver Iron Mining Company had yet to see any profit from Greenway's nearly 10 million cubic yards of moved earth.[24]

The Canisteo was by no means the only stripping operation being conducted on the Mesabi Iron Range. At the time of Greenway's excavation, similar stories were playing out in Hibbing, Biwabik, Chisholm, and other cities on the Mesabi, where tens of millions of cubic yards of overburden were being moved. As in the Canisteo, the steam shovel permitted these unprecedented volumes of material to be moved into the stripping dumps and rock stockpiles that would come to define the skylines of the Mesabi Range.[25]

Although hundreds of acres had been disturbed and tens of millions of tons of earth had been flipped, the conservationist thinkers of the time would have seen them as essentially waste-free operations, having done their duty to preserve the lean ore for future generations. The piles of glacial drift, while not beautiful and difficult to reforest, were considered an inert and necessary by-product of the current generation's duty to develop and use the land's natural resources. Their disposition seemed to be such an afterthought that it is difficult to find any documentation from contemporary mining men relating to the anticipated long-term consequences of creating such mountains of material. The inattention to the visual effects of industry among some political leaders at the time, including Roosevelt's fellow Republicans, is typified by House Speaker Joseph Gurney Cannon's now-famous quote that he would spend "not one cent for scenery."

While Roosevelt's presidency and Greenway's employ on the Canisteo were to end within the next two years, their most lasting contributions to the landscape of northeastern Minnesota were yet to come.

In 1908 and 1909, respectively, Roosevelt signed into creation the Minnesota National Forest (renamed the Chippewa National Forest) and the Superior National Forest, flanking the Mesabi Range to the west and the east. These forests put nearly 1.2 million acres under federal control—an area ten times the size of the Mesabi Iron Range. While the resources of these forests were not preserved from development, conservation principles of scientific management would apply to logging and mining within their boundaries. And while large by Minnesota standards, these forest protections amount to less than 0.5 percent of the land Roosevelt placed under federal protection during his tenure as president.[26]

Having stripped the Canisteo, Greenway's second major challenge was to purify the sandy ore. This problem he tackled with his engineer's training, first studying examples of ore purification from elsewhere in the United States, then running experiments, then iteratively improving his prototypes and designs until he settled on a process that accomplished its objectives. He ultimately built a water-powered purification process in which paddles separated the iron from the silica contained in the Canisteo ore, which became known as washing. It improved the iron concentration from the low to mid-40th percentile up to the mid- to high 50th percentile, thereby transforming it from worthless rock to merchantable iron ore for the Bessemer steelmaking process used at the time. Greenway patented his design as the "Greenway Turbo-washer." The washing works was contained in a building known as the Trout Lake Concentrator—the first of its kind on the Minnesota Iron Range. The sand that was removed from the ore was discharged via flume into Trout Lake.

Late in 1909, the Canisteo district was still a losing prospect, as Boese details: "By that November 1 date, a total of 13,544,948 cubic yards of waste had been removed from the mines in the Canisteo District as a whole at a unit cost of $0.265. The total investment of the United States Steel Corporation in the Canisteo District had reached the imposing sum of $8,186,054.60." However, the Trout Lake Concentrator enabled U.S. Steel to begin to make money on the project. Washed iron ore from the Canisteo, averaging 58 percent iron content, was beginning to find its way to market late in 1909.[27]

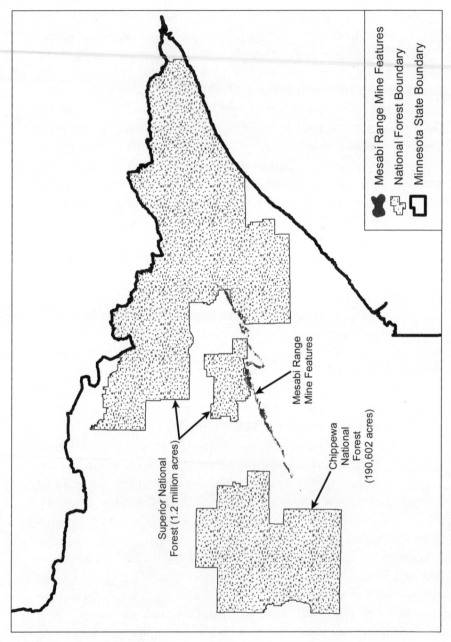

National forests and mine properties of northeastern Minnesota. *Courtesy of Jim Lind*

Mesabi Range Mine Features

National Forest Boundary

Minnesota State Boundary

Superior National
Forest (1.2 million acres)

Mesabi Range
Mine Features

Chippewa
National
Forest
(190,602 acres)

By 1910, it was clear that Greenway's work would permanently change mining on the Iron Range. His efforts set U.S. Steel on a course to mine the Canisteo district for the next seventy years. An article in *Skillings Mining Review* later stated: "Canisteo iron ore concentrate is a well-known trade name in many of the large steel plants of the East, as it is the Canisteo mine which pioneered concentration of iron ore on the Mesabi iron range on a commercial scale." Greenway's invention would remain the largest iron ore concentrator in the world for the next five decades. The new method of steam-powered mining coupled with water-based ore washing swept across the Mesabi Range, opening up whole new reserves and growing mountains of stripped overburden, lean ore, and tailings. At least seventy-eight additional natural iron ore beneficiation plants were built across the range, resulting in such a mining boom that iron ore mining would still be going strong and Greenway would still be remembered as the inventor of beneficiation when its fiftieth anniversary was celebrated across the Mesabi Range—well past Carnegie's predicted twenty-five or forty years.[28]

Celebrating the fiftieth anniversary of Greenway's innovation in iron ore washing along with other beneficiation pioneers, Coleraine, Minnesota, 1957. *Itasca County Historical Society*

Less than a year after startup of the Trout Lake Concentrator, however, Greenway cut ties with U.S. Steel, the Canisteo district, and Minnesota. He moved to Arizona in pursuit of copper for the Calumet and Arizona Mining Company. Meanwhile, Theodore Roosevelt, who left the White House in 1909 having placed over 230 million acres of forestland under scientific management, was enjoying a nine-month safari in Africa.

■ ■ ■

Looking back through the yellowed pane of 110 years of history, the speed with which both Greenway and Roosevelt abandoned their passion projects left me with questions. Having just proved his concepts profitable, how did Greenway feel about the heavily altered Mesabi minescape he helped create? Did Roosevelt's influence ever lead him to blend preservationist principles—not just conservationist ideals—into mining practices? For that matter, did Roosevelt ever foresee a future where conservation and business interests might diverge from the tenuous truce he seemed to have put into practice?

With these questions in mind, on February 13, 2021, my family and I pack our Honda to head into the deserts west of Phoenix in search of the conservationist ideals of John C. Greenway. We have been lured not only by Greenway—who developed Ajo, Arizona, his second "model" mining town, shortly after leaving Coleraine on the Minnesota Iron Range—but also by James and Deborah Fallows, who wrote in the *Atlantic* that Ajo was an oasis of the arts and sustainable agriculture thirty years after the town's New Cornelia mine had closed. In other words, it represented a community-repurposing success story that struck a familiar chord for us while visiting the other side of the country.

I had called ahead to the Ajo Historical Society and reached Mike MacFarlane, president. When I asked whether they had any items related to John C. Greenway, he replied, "You mean John *Campbell* Greenway, the founder of our town of Ajo? Of course!" MacFarlane greeted me by name when I walked through the door of the museum on a Saturday afternoon—the first person to do so since we had come as strangers to Arizona a month and a half earlier. It felt like home in Ajo. Small miners' row houses populated the right-angled streets. The wealth of a century ago radiated from the churches, the Curley School, and the central plaza. And the familiar trapezoid of mine dumps silhouetted the backdrop, except these were green hued, not rusty red like those back home.

From MacFarlane, I learned that Greenway had exceptional grit. During a game for Yale's baseball team, the pitcher noticed the ball that catcher Greenway returned to the mound was getting increasingly bloody. After three innings, the pitcher walked to home plate to confer with Greenway and found him pale, with blood splattered across his face and chest protector. The pitcher, Dutch Carter, recalled, "I then brought him Bliss, the captain, who ordered John to show his hand. The forefinger and thumb were both broken and the bone showed on the thumb and the nail was off the forefinger." The captain ordered Greenway out of the game. When the pitcher asked why he didn't go out earlier, Greenway responded, "You have your job and I have mine."[29]

From José Castillo, a retired copper miner, I heard about Greenway's now-familiar creativity. According to him, Greenway's innovation in Ajo—which had long been a site of artisanal and underground mining—was to unearth the copper deposit and introduce open-pit mining and copper beneficiation methods. Thus, similar to what he had done on the Mesabi, Greenway had brought new classes of mining wastes—lean ore dumps and tailings basins—to this part of Arizona.

Like my hometown of Hibbing, Ajo also had to be moved to enable the progression of mining. When Castillo asked whether I knew who had done the moving here, I politely shrugged while he rolled up his sleeve, pointed to his skin, smiled, and spouted, "Mexicanos!" Castillo's grandfather, an immigrant miner like my own, had been the one to carefully exhume the town graves and reinter them at the new site in 1936. This move had enabled Ajo, like Hibbing, to play an important part in World War II, supplying 25 percent of the copper to the United States during this critical time—and intriguingly, according to Castillo, housing spies and potential saboteurs who aimed to thwart the Allied victory.

But perhaps the most striking fact learned during my visit came from Castillo's story of the Greenway remembrance. He related a family tale of how this remembrance had helped set the Castillos on the path toward financial success and homeownership. Through personal research, Castillo learned that Greenway, upon his untimely death at age fifty-four from a surgery-related blood clot, had left a $250 "remembrance" to each of his mine workers at the time. At first blush, Castillo thought this a nice gesture, but perhaps paltry in comparison to the millions of dollars left to Greenway's wife, Isabella, and their son. However, upon further research, Castillo learned that a house in the Mexican Town portion of Ajo in the year of Greenway's death (1926) could be purchased for $300. Greenway had left the Castillos—and all of his workers at the time—nearly enough to buy a modest house.

In this town that felt like home, I experienced a strong sense of connection to Castillo, a fellow admirer of John C. Greenway and a fellow descendent of the immigrant miners in his employ. So I asked him the question I had come here to find out: "What has been done to reclaim the mine?" He responded, "Nothing. Do you want to know why?" When I nodded, he said, "Because there is still copper in the ground, seven hundred feet down." As in Minnesota, the principles of early twentieth-century conservation had dictated the preservation of Ajo's mineral assets—in the state they were left—for future generations and a day when, possibly, the market or technology would make mining them economical again. In Minnesota, it had been 110 years' wait. Here in Ajo, it was thirty years and counting.

I drove back to my Airbnb rental in Phoenix thinking of Greenway, Roosevelt, and turn-of-the-century American mining towns. Until the death of Barry Goldwater, a statue of John C. Greenway had been one of two representing the state of Arizona in the US Capitol rotunda (and according to Castillo, "Greenway did more for Arizona than Goldwater ever did"). Roosevelt and Greenway—and even Gifford Pinchot—have streets named for them in Phoenix. Clearly, these men are remembered and loved throughout the West as passionate, patriotic, and gritty contributors to the growth of this nation. They were respected by the men who served and worked with them, who, in the case of Greenway, included miners like my great-grandfather, who were able to evade poverty and czarist rule through employment in the mining industry. And while one of them protected hundreds of millions of acres of forestland and the other created mountains of mine waste, they remained friends who were progressive, pro-mining conservationists who believed in American growth and the scientific management of the country's resources. The landscape effects of mining simply didn't register—or, if they did, they were considered a necessary part of progress. Much like our use of toxic chemicals in the post–World War II era or the relinquishment of our personal data to internet companies in the 2000s, it can sometimes take a few decades before the full outcomes of our actions become clear.[30]

. . .

In the days after Roosevelt and Greenway, the conservation movement—and the American political and business interests that espoused it—changed. The next president, Taft, failed to exude the same sort of passion and leadership that drew men like Greenway, and even industrialists like Andrew Carnegie and James J. Hill, into conservation. Rather than progress as one united front led by the most popular president in decades, the conservation movement splintered, frustrating men like Pinchot and losing others who were less passionate.

Across the Mesabi, the process of iron ore washing that Greenway had pioneered spread from the west to the east end of the Iron Range. Along with the expanding mines, a new class of mining waste—tailings—had to be dealt with. At most washing plants, tailings were deposited into the lakes or abandoned mine pits that were used to supply the water needed for the process. Modern researchers estimate that the Trout Lake Concentrator Greenway pioneered eventually deposited 75 million tons of wash ore tailings directly into Trout Lake.[31]

In 1913, water overflowed Wisconsin Steel's washing plant at O'Brien Lake and flooded downstream to Swan Lake, a popular recreation destination near Pengilly, Minnesota. Swan Lake turned red, prompting a group of wealthy cabin owners to file suit against the company. Although the suit was ultimately unsuccessful, its impact led Wisconsin Steel to construct the first upland tailings basin—essentially a settling pond contained within an earthen dam—on the Mesabi Range. Thus, within a few years of Greenway's departure, the open pits, stripping dumps, and lean ore stockpiles of the Mesabi Range were complemented by a fourth type of mine-related landform that would come to dot the landscape.[32]

Within the decade, mining on the Mesabi Range had grown to such a fast rate of output that "every three or four busy years" the volume of excavation equaled "the whole of the work at Panama Isthmus"—Roosevelt's great victory, one of the fifteen modern engineering wonders of the world, which had taken the French and the Americans thirty-four years to finish.[33]

The improvements in mining techniques and mineral processing began to have significant effects on the landscape of the Mesabi Range and its burgeoning communities. With dumping grounds located as close as possible to mines, the pits themselves expanding in size, and the growth of washing plants, the area of undisturbed land in the middle of the range was minimal. In this era before widespread motorized transportation, mine workers had to live within walking distance of the mines. Local historian Marvin Lamppa noted, "Industrial scale mining created problems in the crowded belt of mines, locations and villages, all located within a strip of land six miles long and little more than a mile or two wide. The ever-enlarging pits, with their company offices, machine shops, blacksmith shops, warehouses, draglines, anchor towers, wash plants, railways, trestles, loading pockets, piles of lean ore and expanding dumps of overburden, left little room for towns to grow."[34]

This crowding effect, which left the Iron Range towns surrounded by open-pit mines, stripping dumps, lean ore dumps, and washing plants, became more pronounced after Greenway's departure. In May 1916, a visiting reporter from *Western Magazine*, Edmond L. DeLestry, wrote of the Mesabi Range that "all natural ornamentation disappeared long ago and only huge mounds of debris surround the community, wherever the eye may rest." And although the magazine was "a publication of the west, by the west and for the west"—a region that was presumably then and still today a geographic stronghold of mining support—DeLestry went on to lament that "the mining companies extract from the soil of Minnesota profits running into the many millions of dollars, taking away resources which never can be replaced and leaving in the wake of their operations a barren waste, for the redemption of which absolutely no provision has even been made."[35]

Mining on the scale that occurred on the Mesabi in the early 1900s had seldom been seen elsewhere in the world. In 1922, five of the seven largest mines in the country were on the Mesabi—led by the Hull-Rust of Hibbing, which would later become the world's largest open-pit mine, dubbed the "Grand Canyon of the North."

But as the pits expanded and the mountains of mine waste grew across the Mesabi, some attitudes about resource management began to change. In 1919, Theodore Roosevelt, near the end of his life, lamented the removals of cougars and wolves from national parks (many of which occurred by his own hand), realizing they were, in fact, essential to preserve the land's natural balance of flora and fauna. A young geologist named Sigurd Olson, who in working on the Mesabi Range in 1922 witnessed the outcome of his profession, gave it up, stating, "I went on with my geology work until I discovered that to the end of a geologist's life was mining. Development. And I discovered my interest in the earth, soils and rocks was an intellectual, conservation instinct." He recalled that under the Mesabi skies, he "one day conceived the idea that I should write. I remember it well. It came to me while hiking back to Nashwauk from Keewatin. I shall never forget the glory of that walk." Olson would later move to the forest near Ely (another mining community) and become one of the nation's foremost environmental writers. In 1927, near his new home there, the forest service outlined what is now the Boundary Waters Canoe Area Wilderness within the Superior National Forest.[36]

Even some mining men who had created the minescape grew wistful. Edmund Longyear, who had found and opened so many mines across the Iron Range, wrote in his memoir, *Mesabi Pioneer*, "The forest was gone. In its place was a land of stumps, rocks and mines. I am glad that I saw those beautiful trees before they vanished. For anyone who knows only the modern appearance of the Mesabi Range, it would be impossible to form a true mental picture of the original sites of Hibbing, Buhl, Chisholm, Eveleth and Virginia." And W. C. Agnew, the Mahoning Mine's first superintendent, serving until 1914, recalled, "In those early days, the pine trees reached their tops in the blue of the sky, and for miles upon miles they lay all about us and we dwelt in almost unbroken silence." He later reflected: "It is less than 16 years since we entered upon this forest, but the towering pine is now a curiosity (having been removed by the timber industry) and is replaced by the vast amphitheater of the Mahoning Mine."[37]

Thus, with a growing awareness that something was being lost—even with Roosevelt and Pinchot's conservation in play—the stage was set for a new cast of characters to enter the Iron Range. These entrants brought with them an interest in restoring the lost trees to the Iron Range and examining the spoils of mining for signs of new life. They were the pioneers of mineland reclamation.

CHAPTER 3

The Pioneers

Our program of multiple resource management provides a balance with taconite mining and processing. I have outlined only our work in tailings stabilization, but we are also working in forest management, which includes a wildlife program. These programs were underway well in advance of the enthusiasm of modern environmentalists and present legislation. They show what can be accomplished by the determined cooperation of owners and management to keep mining compatible with the environment.

Sam Dickinson, "Revegetation of Taconite Tailings,"
Mining Congress Journal, October 1972

In days after the world wars—as the nation venerated its heroes who had prevailed against uncertain odds—the contributions of the Mesabi lands were not forgotten. Lakes had been filled or drained for mining, and billions of tons of earth had been moved to create the nation's war machine. Molecules of Minnesota iron had rolled in tank treads across the battlefields of Europe and bobbed inside the hulls of US battleships that sailed the South Pacific. By the end of World War II, Minnesota was producing 68 percent of the nation's iron ore. One mine alone—the combined Hull-Rust-Mahoning pit in Hibbing—was hailed as having earned "a hero's reputation" by giving up "more iron ore than the entire production of any foreign country." What would Germany have given for the Hull-Rust and its 100 million tons of iron ore? It was later hailed as a national landmark and compared in scale to the Empire State Building. Even today, Iron Rangers frequently celebrate their region's contribution to winning the world wars and building the foundation of American prosperity.[1]

The postwar presidents, unlike their predecessor Theodore Roosevelt, took notice of the Mesabi and its riches. When Harry S. Truman

At the edge of the Hull-Rust-Mahoning pit, the largest open-pit iron mine in the world, near Hibbing, 1942. *Library of Congress*

met Hibbingite John Galob in 1947, the president was reported to have said, "I know Hibbing; that's where the high school has gold doorknobs." Hibbing's reputation as "the Richest Little Village in the World," located in the center of the United States' largest and richest iron ore deposit, was spreading. During this era, Ayn Rand made the

Mesabi Range the homeland of her fictional capitalist hero in *Atlas Shrugged*: Hank Rearden, after working a double shift in the iron ore mines, "wanted nothing on earth except to lie down and fall asleep right there, on the mine ledge." By age thirty, Hank owned the mine, and as the book progresses, he is one of the only people left in society who knows how to work and think for himself. Another self-made Iron Range millionaire was depicted in the 1944 MGM movie *An American Romance* (originally filmed under the working title *The Magic Land*), in which an immigrant miner eventually works his way up to become an automobile and then an airplane manufacturer. Bootstrapping mining millionaires also existed in real life: William Boeing's family investments in Mesabi timberlands and iron mines provided the seed money for Boeing aircraft company. From the mid-1940s through the early 1950s, the Iron Range was arcing toward its zenith of both mine production (which occurred in 1953) and notoriety.[2]

Along with the acknowledgment of the Mesabi's great productivity and wealth-generating potential came a slow-dawning and quiet recognition that the scale of land disturbance was also colossal. As mining continued to advance and more stripping dumps, rock stockpiles, and tailings basins were left behind, a new sort of pioneer emerged on the Iron Range. These new pioneers weren't geologists, miners, or land speculators seeking the Mesabi's mineral wealth—many, in fact, were just kids and college students in search of a different kind of value. They sought to understand and restore the ecological, recreational, and social worth of the lands that had given us all so much iron.

■ ■ ■

In the middle of 1952, a young biologist climbed the mine dumps of the Mesabi. This was a summer before rock and roll got its name. Elvis had not yet debuted a single; rather, the hit parade was occupied by the likes of Rosemary Clooney and Eddie Fisher. Baby boomers were just being born. Later that year, the Great Smog of London would kill some twelve thousand English, but the early-summer weather on the Iron Range was unusually dry and hot.

Gilbert Arthur Leisman would have known England from his time as a navigator in the army air corps, when he had been called up to serve in the European theater. Now he was twenty-eight years old, a veteran, and a master's student at the University of Minnesota. His fascination with nature had led him to botany, and his interest in the naturally occurring revegetation of industrial lands led him to the Iron Range. He was studying how nature—when left to its own devices—repopulates disturbed land with plants and begins re-creating topsoil. He used words such as *ecosystem* and *waste land* but generally avoided hyperbole regarding the subject of his studies. His first summer of fieldwork was spent "in general reconnaissance of the Mesabi Range to ascertain the extent and possibilities of various spoil [waste rock] piles." What he failed to mention in his peer-reviewed journal articles was that his 1952 field season was interrupted by his marriage on June 21—which is, perhaps, why his full first year in the field was spent performing only "general reconnaissance." He would go on to become a PhD graduate of the University of Minnesota and a long-serving, well-published professor of paleobotany at the Kansas State Teachers College. But in summer 1952, the newlywed was trying to keep his footing as he scrambled up the steep, red angles of repose that constituted the Mesabi mine dumps.[3]

At the time, there had been relatively few studies on the natural revegetation of industrially disturbed properties. Some states required areas that had been strip-mined for coal be replanted, but there were no such requirements for Minnesota's iron mines. Therefore, the Mesabi Range, with more than five decades of continuous steam and later electric shovel–enabled, large-scale mining, proved to be a perfect laboratory to study naturally occurring mineland revegetation. Leisman roamed the range looking for similar stockpiles of varying ages to compare natural regrowth of vegetation across time. He ultimately selected ten, ranging from two to fifty-one years old—representing every decade of natural growth—and covering the portion of the Iron Range from Virginia to Buhl to Kinney. He studied six stripping dumps, the flipped pancakes of Iron Range overburden—which he called "man-made hills and plateaus of glacial till"—and four "lean ore" stockpiles, which, at the time, were primarily composed of taconite (not yet considered an iron ore because of its relatively low iron content). Some of the towns he referenced in his 1957 publication, such as Cooley, no longer exist, having given way to mine expansion. But nine of the ten overburden and lean ore stockpiles are still there.[4]

What Leisman found in a two-year-old stripping pile was that small patches of *Trifolium repens*, white clover, were growing in the surface

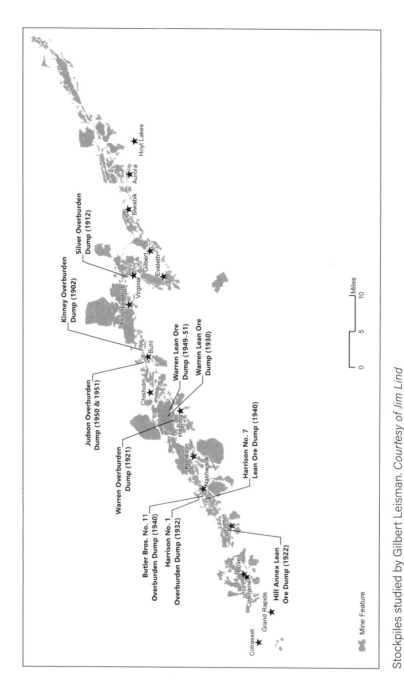

Stockpiles studied by Gilbert Leisman. *Courtesy of Jim Lind*

depressions of the otherwise barren landscape. These small depressions collected both windblown seeds and, most importantly, moisture. In other words, in what many people might write off as weeds, Leisman saw nature getting a foothold. On a decades-older stripping dump, he noted a dozen more species of weeds and the first evidence of trees— poplar and pin cherry—whose fallen leaves had accumulated to create about half an inch of organic topsoil. This trend continued decade after decade such that a thirty-two-year-old stripping pile had a "very uniform woodland community, composed of aspen, poplar, birch and pin cherry" as well as occasional seedlings of balsam fir. Other stripping piles developed as grassland communities, which set Leisman to wondering why some piles favored grasslands over trees. But the overall trend was clear: the cover, both herbaceous (weeds, grasses, forbs, and so on) and woody (trees), was naturally moving into—and developing topsoil on—the stripping dumps. He also found that "animals, especially birds, play an important part" in spreading the first seeds and thus in the development of early vegetation on the bare areas. In light of his findings, it should be no surprise that the Winston and Dear Dump surrounding my house on Pill Hill is now occupied by wild fruit trees, aspen, and dozens of species of ground-covering herbs.[5]

Leisman extensively studied the repopulation of stripping dumps by aspens. In northern Minnesota today, these species are regarded as commonplace, trashy even. They are neither as vibrant nor as good for burning as maple or oak. They are stark in winter compared to our evergreens. And while their tops seem to snap off easily in a strong wind, they are strangely difficult to mill as shop wood. Despite their role in creating seventy-five-foot-tall forested canopies over previously treeless lands, and despite the quaking wind song of their leaves in a summer breeze, aspen in Minnesota have never achieved the reverence they receive in Colorado. But Leisman found them to be a hugely important species in terms of mine-dump reforestation, with one unique reproductive property.[6]

Long before Dolly the sheep was cloned in a laboratory, aspen naturally produced clones. The method is root suckering: new tree trunks shoot upward from the same shallow root system of the original tree. In this way, one parent tree can produce dozens of genetically identical trunks. These clone colonies, which appear to be many trees, are in truth just a single organism. The US Forest Service describes the uniqueness, like a fingerprint, of each individual clone colony, which "can be distinguished from those of a neighboring clone often by a variety of traits such as leaf shape and size, bark character, branching habit, resistance to disease and air pollution, sex, time of flushing, and autumn leaf color."[7]

A single clone can stretch over dozens of acres. The largest known aspen clone, named Pando, is an individual male covering more than a hundred acres of the Fishlake National Forest in south-central Utah. It is speculated to be one of the oldest living organisms on earth. In the western United States, aspen colonies that are five to ten thousand years old are common. The change in climate has made it difficult for aspen to reproduce by seed in the West, which in part explains their prized nature in places such as Colorado, despite being "the most widely dis-tributed tree in North America with close kin on three continents." In these barren western environments, the aspen sustain themselves and spread by root suckering more so than seed reproduction. In *The Overstory*, author Richard Powers poetically describes Pando and his root suckers "roving around the hills and gullies in a ten-millennium search for a female quaking giant to fertilize."[8]

On barren mine dumps, where the shade-intolerant aspen are given full exposure to the sun, root suckering can happen almost unimpeded by competition, resulting in trees that grow three to four feet in a season. With this remarkable reproductive feature, Leisman found that natural reforestation of overburden piles could occur if just a few seedling trees took root. At one seedling every fifty feet, Leisman calculated that in just twenty-five years root suckers could produce a solid tree cover on a barren stripping dump—and in fifty years a complete canopy.

It is easy to distinguish one aspen clone from another with the na-ked eye. Each typically forms the shape of an arched eyebrow—like the roof of Minnesota's old Metrodome—with the parent tree somewhere near the middle and smaller, newer clones radiating outward with a slight preference for the south (because the south-expanding clones suffer less shading effect than those on the north). This creates some-thing of a dome with a southern tail, not unlike the half-teardrop shape imprinted across the Mesabi in the drumlins (rounded, ovular hillocks left behind by glaciers) like Maple Hill in Hibbing or the long field of drumlins one drives across near Toimi on Highway 16. Under these teardrop-shaped aspen clone domes, the more highly prized evergreens and maples take root in a process that foresters call succession.

Leisman also studied the coarse rock stockpiles composed of golf ball– to softball-sized taconite chunks, which he described, in botanist's terms, as "extremely coarse-textured substrate." What he found was that—over a thirty-one-year period—a surprising number of trees took root upon the rock stockpiles. While these stockpiles remained utterly bare two years after deposition, by their third decade they had naturally regrown a handful of trees over four feet in height as well as a few ground-cover species. Why would trees grow on what is essentially bare rock? After much study, Leisman concluded that the nooks and crannies in the rock could capture both moisture and the wind-, bird-, and animal-spread seeds and ultimately proved "probably quite favorable for seed germination."[9]

Whereas aspen seemed to dominate overburden piles, Leisman observed pines on the rock stockpiles, and he especially noted that "paper birch played a somewhat more prominent role" in the reforestation of lean ore stockpiles. I have personally confirmed birch's preference for lean ore on my own home stockpile of Pill Hill, where the seventy acres of forest underlaid by overburden contains only five birch trees, but the small two-acre corner where the lean ore is exposed is populated by hundreds or thousands of individual *Betula papyrifera* trunks. This phenomenon is also on display along US Highway 169, where the white, angular trunks and bright green (or yellow, depending on the season) leaves of paper birch are highly visible against the rusty-red and gray rock stockpiles. This species, which is rumored to be dying

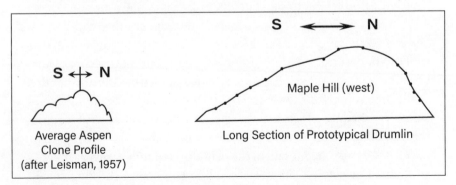

Shapes of aspen clones and drumlins. *Courtesy of Louise Lundin*

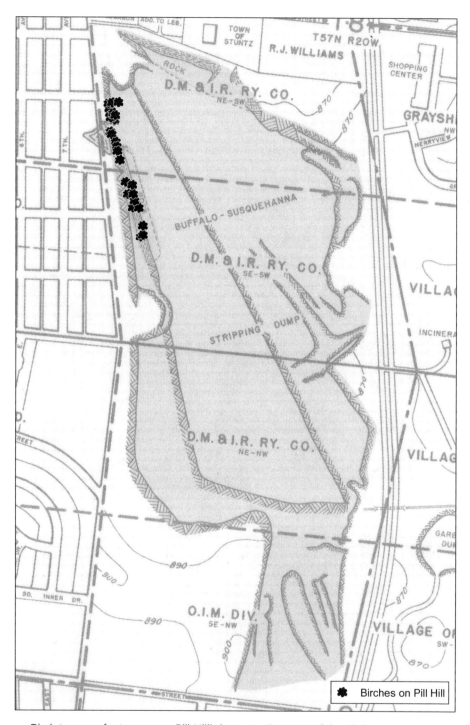

Birch trees prefer to grow on Pill Hill's lean ore. *Courtesy of Jim Lind*

in other parts of Minnesota, somehow seems to thrive upon the rock piles of the Iron Range. Birch is classically a pioneer species that under natural circumstances may give way to succession; however, on the rough-hewn rock substrate, its "pioneering" goes on decade after decade.[10]

This phenomenon, in which natural succession has slowed into a

Birches growing on rough rock. *Author's collection*

sort of protracted youth, has been observed on disturbed, postindustrial properties elsewhere in the world. Author Cal Flyn describes English brownfields where "old concrete and tarmac hinder succession," thus creating an enduring habitat for species that might otherwise be ephemeral. Where these species are rare or endangered, the disturbed sites can become "refugia for wildlife" that are sometimes noted for their incredible diversity of rare species, "only equaled by that of some ancient woodlands"; "the spontaneous ecosystems that have grown up on wasteland, unsupported . . . are a new form of wilderness worth preserving in their own right." She concludes: "Indeed, there had been a sea change in how post-industrial, or other 'anthropogenic' sites are perceived and valued in recent years. Some of the most exciting developments in ecology and conservation have been in the study of landscapes deeply impacted by human activity."[11]

Such is the case on some of Minnesota's minelands, where a tiny, endangered species of the genus *Botrychium*—better known as grape fern or moonwort—has been identified. The plant is just an inch or two tall, is visible only in the early summer, and produces just one leaf per year. According to the Minnesota Department of Natural Resources, its presence is disturbance dependent, meaning that as open habitats turn into forests, they become unsuitable for the moonwort. But minelands, which are heavily disturbed and slow to succeed into mature forests, are havens for these rare plants—at least until natural succession occurs and *Botrychium* need some freshly disturbed areas to call home.

■ ■ ■

Intrigued by these findings, in June 2022—nearly seventy years after Leisman's original work—I took it upon myself to revisit some of the stockpiles he studied. I wanted to see how Mother Nature had further revegetated them with seven more decades of behind-the-scenes work.

It was a damp but clear Sunday evening when I pulled into Kinney, Minnesota, a mining location town built near the now abandoned Kinney Mine. It has just twelve blocks of houses and one commercial establishment (a bar), but, in a manner typical of the early wealth of the small Iron Range communities, is served by a host of public amenities, such as a library, a city park, potable water, and a small township garage to house snowplows. Kinney achieved a modicum of fame in the 1970s when all efforts to attract public investments in its aging infrastructure failed and the town council decided the best course of action was to secede from the Union, declare war on the United States, get bombed, and then finally get rebuilt for free (a political strategy known as the "Republic of Kinney"). Today, Kinney is still part of the United States and serves as a small bedroom community. The Kinney Mine closed in 1937 and is refilled with groundwater. The northern shore of the pit lake is fitted with a boat launch and a public fishing pier to catch stocked rainbow trout from the clear, clean waters.

When Leisman visited Kinney in 1953, the overburden piles south of the townsite were the oldest he studied, having been placed fifty-one years earlier, in 1902. At the time, these stripping piles supported a "uniform woodland community" of *Populus tremuloides* (quaking aspen) with occasional poplar and paper birch scattered in. There were no conifers or shade-tolerant deciduous trees.

As I tried to find the 120-year-old stripping dump, I had to check my map twice to make sure I was actually looking at a mine stockpile. It was late in the evening when I finally parked my car beside the dump, which now belonged to the public as a tax-forfeited property. I had to don my waterproof hiking boots to cross the roadside ditch growing up with cattails, buttercups, and daisies to reach the dry ground. Once I made it up onto the pile proper, I could finally recognize the prototypical landform of a trestle dump—the flat, scallop-shell top formed when the rail lines had fanned out to dump their loads of glacial drift and the steep, angle-of-repose side slopes dappled with random large boulders that looked like glacial erratics but had actually just been dumped from above by railcar. Although the understructure of the landform was unchanged from Leisman's time, the similarity ended there. Where Leisman had seen no conifers, now the balsam firs were so thick as to create premature dusk in the understory on the western slope. Atop the pile, where Leisman saw no shade-tolerant species, there was now an understory of deciduous sugar maple, which I encountered every two or three strides, and even one small oak. In the places where aspen and poplar remained, they were either so huge that my arms when wrapped around them touched only at the wrists or fallen over and rotting back into the earth, opening up the canopy for the ascension of young hardwoods beneath.

Conifers growing on the Kinney stockpile. *Author's collection*

Sugar maples growing on the Kinney stockpile. *Author's collection*

Somewhere nearby, a whitetail snorted behind the thick foliage in an effort to drive me off this human-made mountain, but I was already circling back to my car on account of the whining fog of mosquitoes that enveloped me. Back on the road, I saw a mole scamper across the pavement as well as the crushed body of a less lucky painted turtle, hunched over by a large bird whose plumage was difficult to discern in the darkened dusk. All in all, the stripping dump seemed to my layperson's eyes to be as bountiful and thick as any forest I'd walked through in the Boundary Waters. While my visit had been short and un-scientific—hardly a robust survey of the biodiversity of species or their overall health—I could say that, at 120 years of age, this overburden pile wasn't easily distinguishable from a natural drumlin in terms of its vitality. To top off the experience, on my sunset drive home, a three-hundred-pound lumbering black bear crossed Highway 169 before me. I concluded that, on this one stripping dump, at least, no further human endeavor would be required or should be attempted to improve upon nature's work. It seemed to have made a full comeback.

The lean ore dumps were less simple to assess. Two of the four existed on private land. Another had been experimentally planted in the 1980s (erasing the opportunity for an unadulterated reassessment of the growth). And the fourth had been mined through, but wasted because it contained ore that was too oxidized to be properly processed in today's mines. The best I could do was assess—from the legal distance of the public right-of-way—a pile of similar vintage to one of Leisman's lean ore stockpiles. Both were remnant rock piles from a natural ore mine with a Greenway-inspired concentrator located near Nashwauk. They had been idled for sixty years, and the rock sizes were three to six inches in diameter—Leisman's "extremely coarse-textured substrate." In this instance, Leisman's findings were very similar to what I witnessed with my own eyes. Paper birch—small, individual trees—appeared to be thriving, especially near the base of the rock pile. Quaking aspen, bigtooth aspen, and balsam poplar were present in smaller clumps, with trunks that were stunted and cankered. And a few scattered red pine and jack pine existed, especially near the top of the dump. But the ground cover was sparse, and the rate of soil creation appeared to be either slow or nonexistent. Much of the steep red slope was as bare and straight as the Great Pyramid of Giza. In general, the lean natural ore stockpiles—which were never intended to be permanent but had now been present upon the Mesabi landscape for so many decades that their trademark appearance can be seen on business logos (like West Range Storage) and city crests (like Gilbert's)— were still visibly angular, red rock landforms broken up by occasional birch and evergreens.

After seeing these stockpiles with my own eyes, it was clear to me that Leisman's early work had accurately projected how the Mesabi minelands would naturally recover from disturbance. He shined a light on how the often-underappreciated pioneer species, such as aspen and birch, work to repopulate the minelands by building soil and making the terrain suitable for successional species to thrive—a process that had been carried out to completion on the 120-year-old stripping dump near Kinney but was still very early in progress on the lean ore stockpiles. To reuse the term applied to the Hull-Rust in World War II, it seems to me that these scrappy volunteers deserve not our disdain but rather a hero's reputation in the natural reforestation of the Mesabi stockpiles.

■ ■ ■

While some students were focused on botany, other Iron Range youth of the 1930s through 1950s had more thrilling uses for the mine dumps that enveloped our communities. In the summer of 2021, I walked into Realife Cooperative in Hibbing to talk with ninety-four-year-old John Dougherty about his childhood on the Mesabi. When I asked how he felt about growing up among the mine dumps of Hibbing, Dougherty replied, "Great! They were our playgrounds. Pill Hill at that time was simply known to us as 'the Dumps.' That is where I got into skiing."

Dougherty went on to describe how he—and other neighborhood kids—grouped together to downhill ski and, eventually, ski jump, on the Dumps. They had three jumps: Little Speedway, Big Speedway, and the Scaffle (a misnomer for the Scaffold). He explained, "This is where I learned the rudiments of jumping. We started on Little Speedway. Graduated to Big Speedway. Eventually you went to the Scaffle." Once you climbed to the icy top of the Scaffle, he said, there was the next level of challenge: "You knew you had arrived when you could clear the knoll. When you can clear the knoll, you were a member of the elite."

I immediately thought of all the runs down the Scaffle where kids *did not* clear the knoll and, therefore, would have had no option whatsoever for a safe landing. This is when I noticed Dougherty's knee, exposed in summer shorts only a couple of feet away from my own. His, like mine, showed the trademark scar of surgical reconstruction. The matching scars were like a set of railroad tracks running up to a trestle dump. "Ski injury," Dougherty confirmed, to which I retorted, "Volleyball."

I eventually got around to asking the question that interested me the most: Who created this recreational life for children on the mine dumps? "We built our own ski jumps," he said. And who gave them the skis? "We made them from barrel staves." According to Dougherty, eventually the city got on board and built a toboggan run to complement the ski runs and jumps the kids had built. Other cities, like Virginia, followed suit. But clearly, the children—motivated by their drive

for adventure and desire to enliven the monotonous Mesabi winters—
had simply created their own rough-and-tumble recreational opportunities on the Iron Range minescape. Whether they owned the land or
not, they laid claim—or "reclaimed"—the mine dumps for themselves
and used the steep, angle-of-repose rock dumps to turn the Dumps
into their own miniature Zermatt. Dougherty would draw on this up-
bringing and ski training later when he visited the real Alps. He told
me he went on to ski sixty-five different venues in the United States,
Canada, and Europe. The culmination of his skiing avocation was a
trip to the Matterhorn, where he took a tram up to the top, then skied
down the mountain and "straight into a bierstube." After refreshment,
he took the tram back up and skied an easy blue run back to his lodg-
ing in Italy. During this ski trip of a lifetime, Dougherty said, "I was
thinking, this little bowlegged kid from Hibbing who started skiing
on the Dumps ended up skiing around the Matterhorn." While he was
focused on his skiing, he didn't know then that he and his friends were
pioneering the recreational use of the Mesabi's mined landscape.

■ ■ ■

Another boy, one who lived less than a block from the ski jumps Dougherty and his friends built, was surely recreating on the Dumps at this
time, although he wouldn't speak of it publicly for decades. I like to
imagine he was remembering a toboggan ride with Echo Star Helstrom (whom the San Jose *Mercury News* called "Hibbing's Brigitte
Bardot") when he penned the line "What's a sweetheart like you doing
in a dump like this?" Indeed, she was later suspected to be the "Girl
from the North Country." But during his rise to popularity, Bob Dylan
veiled his Hibbing roots in obscurity, claiming one day that he grew
up in the South and another that he had had an itinerant childhood. In
those early years of fame, he was famously dodgy, and it would have
been hard to know what he thought of his Iron Range homeland.[12]

But Dylan eventually admitted, "I was born and raised in iron ore
country—where you could breathe it and smell it every day." Helstrom
likewise described her home to Dylan biographer Anthony Scaduto as
"all woods and iron ore dumps" and recounted how she and her friends
used to "take our bikes up to the dumps. We liked to go climb on them."
Dylan also said his "brains and feeling have come from there," where
"the earth . . . is unusual, filled with ore." Referring to North Hibbing,

which by that time had been abandoned when the whole town moved south to accommodate mining, he said, "It was as though the rains of wartime had left the land bombed-out an' shattered."[13]

But Bob's house, along with the ski jump and toboggan run, was in South Hibbing, "where everybody came t' start their town again." He later expounded on the "birch trees, open pit mines, bears and wolves" and—I imagine remembering the Hibbing winters on the Dumps— recalled that "the air is raw." He once told *Rolling Stone* that "the forests are thick and the landscape is brutal" but that, nevertheless, "the sky is still blue up there. It's still pretty untarnished. It's still off the beaten path."[14]

Of his middle-class childhood, Dylan wrote that Hibbing "was not a rich town or a poor town; everybody had pretty much the same thing, and the very wealthy people, the ones who owned the mines . . . lived thousands of miles away." John Dougherty told a grittier story. In a humble manner brimming with gratitude, he said:

> I'll tell you how lucky I was to have good food and [a] warm place to sleep—when some didn't. My mother was a wonderful cook. We were so lucky. We had some very poor neighbors who were short of food. Particularly one family who lived across the alley. My mother would set out food on a regular basis. Three kids at home, no mother, being raised by their father. He used to get a screen, like a screen from a door, and prop it up. And then he would sprinkle some bread crumbs around. He would pull the string, knock the birds, and they would have that for dinner. Trapped songbirds.

Clearly, growing up in the "Richest Little Village"—a place Dougherty recalled also being known as the "Razzle Dazzle Village"—meant little to the economic prospects of the middle- and working-class folks who operated the mines and grew up in the minescape.[15]

In 1956, Bob Dylan was famously pulled off stage by the Hibbing High School principal for raucously performing a Little Richard cover during a talent show. He stuck around until he graduated in 1959, but then, in his own words, "I ran, an' kept runnin'." Though the youth of the Iron Range were innovating on the landscape, Hibbing was not yet ready for Dylan's style of musical creativity. When I asked Dougherty, who conducted Bob Dylan's father's funeral, what he thought of

Dylan at that time, he said simply, "talented a-hole"—a snub directed at Dylan's early dismissal of his hometown. But years later, the musician made amends. Speaking of the Iron Range, he said, "there's a magnetic attraction there" and "a great spiritual quality throughout the Midwest. Very subtle, very strong, and that is where I grew up." It sounded like long-overdue love of homeland coming from the Voice of a Generation.[16]

■ ■ ■

Just a year after Bob Dylan left, John F. Kennedy came to Hibbing for a political rally. John Dougherty watched as Kennedy walked into the Memorial Building arena, but Dougherty did not attend the speech because he "had no money for political donations." In the same way that JFK later stood with the citizens of Berlin during the airlift crisis—famously declaring, "Ich bin ein Berliner"—he embodied the importance of Minnesota iron ore to America when he said, "This is the Iron Range, and the Iron Range is the power of the United States."[17]

At that time, the supply of iron ore was our source of power. But that was beginning to change. By 1960, less than half of the nation's iron ore needs were being filled by the Lake Superior region, a sharp decline from its peak. The washable natural ores here were nearing depletion, and a new technology for iron mining was on the rise.[18]

■ ■ ■

On September 30, 1957, the Iron Range was changed forever. On that day, the first boatload of Mesabi taconite pellets was shipped from the Erie Mine. Erie, owned by Pickands Mather & Company, was a new type of mining venture for the Iron Range. Like John C. Greenway's Turbo-washer that allowed the sandy iron of the western Mesabi to be transformed from waste rock into ore more than four decades earlier, the production of iron ore pellets from low-grade taconite, the "mother rock of the Mesabi," changed the value proposition for vast swaths of the Iron Range. The iron percentage in the taconite mined by Erie that year was just 22.52 percent—less than half of what miners considered to be ore a half century earlier. Billions of tons of rock that had previously been deemed unusable was now economical to mine. Walter Havighurst exuberantly states the significance of this innovation to the future of the Iron Range in *Vein of Iron: The Pickands Mather*

Story: "If iron ore could be profitably produced from the mother rock, the industry could go on with no foreseeable ending."[19]

The mining engineer whose life's work had made the economical mining and processing of taconite possible, Edward W. Davis of the University of Minnesota, used a different analogy. Likening natural iron ore to "plums" and taconite to "pudding," he said one may quickly exhaust the prized plums, but of the surrounding pudding, "we shall still be 'eating' for the next five or six hundred years."[20]

This innovation opened the third wave of mining on the Iron Range. Taconite purification had begun as a puzzlement—in Havighurst's words, it was a "heartbreaker"—but thirty years of experimentation led by Davis and others had elevated it to the status of "life-giver." Taconite is still giving life to the communities of the Mesabi range as the only type of iron mining ongoing here today, and it is estimated to have the resources to continue for at least another century.

The construction and operation of the Erie Mine was a colossal project involving, as Havighurst puts it, large risks, huge financing, and complex technology. Not only that: the "largest iron-mining venture in history required mountains of material and equipment and thousands of workmen." Shortly after the mountains of material and equipment were installed and operating, the Erie Mine began creating mountains of its own.

The simple mathematical fact is that ore made from lower-grade rocks will produce more waste than higher-grade ores—50 percent pure ore is half waste, whereas 25 percent pure ore is three-quarters waste. The lower-grade ores have more impurities that need to be removed to produce an economical metallic concentrate. At Erie, it took three tons of taconite to produce one ton of iron pellets for the blast furnace. The remaining two tons of material was separated—through an elaborate process of grinding, magnetic separation, and purification designed by Davis—and deposited as waste tailings. These tailings consisted of just the waste products from the *ore* and did not include the overburden and waste rock produced from stripping.

According to Pamela Koch's history of Erie Mining Company, "To achieve Erie's rated capacity of 7.5 million tons of pellets per year, it was necessary to mine 37 million tons of material, consisting of 24 million tons of taconite ore and 13 million tons of stripping." That meant

Three tons of taconite yields one ton of pellets for the blast furnace and two tons of tailings, highlighting the importance of mine waste management in taconite mining and processing. *Minnesota Discovery Center*

that over 80 percent of the material being moved by Erie was waste material of one sort or another. And Erie was moving a lot of material: "Each month . . . Erie moved more material than the average large Mesabi Range natural ore mine did in a year." The productivity that steam shovel mining had brought to the Iron Range in the early 1900s had now been eclipsed by the material movement involved in taconite mining. Similar expansions in the footprint of mine waste were occurring elsewhere in the Lake Superior region. It was around this time that a mining company purchased my great-grandfather Matti's farmstead in Michigan—which he had sailed from Finland and worked under Greenway to secure—and covered it with a waste rock stockpile.

The Kero family farm near Palmer, Michigan, was later covered by a mining stockpile. Shown here are the author's great-grandfather and uncle. *Author's collection*

It is possible the rocks that were placed on the family farmstead were dug by my grandfather, Matt Kero Jr., who was a mine shovel operator at the time.[21]

The huge quantities of material movement and high percentage of unusable rocks and unconsolidated material meant that taconite mining was more about waste management than it was about hitting pay-dirt. Tailings basins required dedicated staff and were often the size of the small cities their operators and engineers went home to after their shifts ended. Taconite tailings were finer than natural ore tailings—like talcum powder in comparison to coffee grounds—and consequently could create a dust problem if they were dry or frozen and affected by high winds. Furthermore, taconite tailings were utterly devoid of organic material, making them essentially barren with respect to the possibility of becoming naturally vegetated by grasses or trees.

The new problems created by taconite waste fell to a thirty-three-year-old forester living in Aurora, Minnesota. Sporting horn-rimmed glasses and a military physique, the young Sam Dickinson was accustomed to thorny problems. As a boy, he had managed the academic and social challenges of splitting his school year between the public schools of Sparta, Wisconsin (his family home), and Lutsen, Minnesota (the location of his family's cabin on Caribou Lake). During and after World War II, he had served on medium-sized landing vessels for the US Navy in the Pacific, achieving the rank of lieutenant, junior grade, and serving several tours near Japan. He had studied his way through undergraduate and master's degrees in forestry from the University of Minnesota and then Syracuse University. Now he was facing a challenge that would require all of his academic, social, and physical strengths to surmount: creating a way to grow things on unprecedented volumes of waste material for a type of iron mine that had not existed just three years earlier.

Dickinson's job entailed more than just the revegetation of waste materials. In fact, when he was first hired in 1954, Erie Mining wasn't even contemplating mineland revegetation when he was assigned to survey the landscape and oversee the logging of the forest on the taconite lands the company targeted. In those days, surveying was a much more physically demanding job, requiring its practitioners to hand-cut property lines through the woods and measure distances with hundred-foot-long chains. Surveyors needed to be handy with a saw,

Sam Dickinson, a forester for Erie Mining, assessing revegetation of a taconite tailings dam, 1972. This work was performed years before mandatory mineland reclamation. *Courtesy of Kent Dickinson*

chains, and a compass; good at geometry; and comfortable traversing all types of lands in any kind of weather. Dickinson also needed to coordinate and manage the work of hardscrabble loggers, making sure they were meeting schedules and clearing the correct locations, all while generating a revenue stream for Pickands Mather & Company to invest in the construction of the Erie Mine and processing plant.

Dickinson loved both the academic and outdoor rigors of his job. His coworker and fellow forester Dave Youngman, whom Dickinson had personally recruited away from a possible career with the US Forest Service, described him as "one of the best foresters I've ever met." He was a great experimenter and was known for his thoroughness and attention to detail. Youngman said that any job assigned to Dickinson "was a job that was done extremely well." Dickinson, like most northeastern Minnesotans, also loved to hunt and fish. He gave back to his community by coaching youth hockey. And he was especially known for his love of making maple syrup—which in northern Minnesota requires the hardiness of someone who can carry bucketloads of sap through a crusted-over snow blanket in spring, sometimes unpredictably breaking through the icy surface to sink up to one's waist, just to provide enough raw material to make one pint of syrup.

Dickinson exceeded the expectations put upon him. He brought to Pickands Mather a concept he had learned in forestry school called multiple resource management. The idea was that any tract of land had multiple resources upon it and should be managed to provide the greatest benefit to the largest number of people. Under multiple resource management, trees atop a future mining area would not simply be bladed off by bulldozers (as they were in other mining operations at the time) but carefully harvested to provide timber for construction, papermaking, and the townspeople's use. Pickands Mather allowed the public to hunt and pick berries on company lands. In the fall and early winter, it issued firewood and Christmas tree permits to the local residents free of charge. And fishing and swimming were allowed in company water reservoirs.

It was novel for a mining company to practice resource management that valued not only the minerals but also the timber, recreation, and public relations assets of the land under its control. According to Youngman, Dickinson had seen multiple resource management in use by the US Forest Service. And, in Youngman's words, Dickinson "didn't see any reason why we couldn't have this same kind of philosophy for company lands."

In a 1971 talk delivered in Duluth, Dickinson summarized multiple resource management as "fully utilizing its lands to provide for all compatible uses." The resources for which Erie managed its lands included minerals, timber, water, wildlife, and recreational uses. It is important to note that only two of these—minerals and timber—had the potential to bolster Erie's bottom line. Despite this, in his talk Dickinson stated that "the ecological requirements of fish and wildlife are the *least* flexible aspects of the Multiple Resource Management Plan" and that "*all* land management decisions are given close scrutiny with regard to recreational potential" (emphasis added). In other words, the non-revenue-generating activities were given as much consideration as those that generated revenue.[22]

The employment of multiple resource management made Erie Mining a public-friendly company decades before today's notions of social responsibility. Youngman recalled, "Most of the mines up here wanted to keep everybody off of their property. Erie Mining Company was not like that. If there was an operating area or other sensitive locations where they didn't want hunters or bikers or people walking, they would post it. But all of the other areas, which was more than sixty thousand acres, was open to the public." He added that "people would not go into posted areas because they had other places to go" and that liability for public use of company lands was not a concern because "there is a recreational law in Minnesota that protects the company. It worked out just fine. They wouldn't charge anybody for any of this access, firewood, or Christmas trees, and it worked out just fine."

Erie's philosophy of using its resources to provide the greatest benefit to the largest number of people created some unique opportunities for a mining company. In 1970, Erie offered to provide Minnesota-grown Christmas trees for the annual Pageant of Peace outside the White House in Washington, DC. Dickinson and Youngman walked their future mining areas looking for fifty-two well-balanced and beautiful twelve-foot-tall balsam fir trees. Balsam standing in a copse are sparse trees with horizontal boughs—not exactly textbook Christmas trees—and so this task required sturdy legs and a discriminating eye. Dickinson created a "viewing card" with a tree-shaped cutout to help select trees that would be full and uniform in appearance. After they chose the trees, Erie's general shops—a fully equipped fabrication and maintenance shop—made a mesh-wrapping device to ready the balsams for their journey. In a special event held in November, the mesh-wrapped Christmas trees were shipped from the mine to Taconite Harbor, where they traveled on the ore boat *Herbert C. Jackson* to Cleveland, then on an eighteen-wheel semitruck to Washington, DC. The Minnesota balsams on display provided a little Christmas cheer for the nation's lawmakers, but more importantly, they highlighted the kind of public goodwill exhibited by Pickands Mather & Company and the Erie Mine.

Dickinson applied the philosophy of multiple resource management most ardently to reclaiming lands disturbed by Erie's mining activities—its massive pits, overburden piles, rock stockpiles, and tailings basin. He took the initiative to arrange a pivotal meeting with his boss, the company's general manager. In an era when no laws required the shaping or reforestation of mined lands in Minnesota, he drove the

Balsam from Erie Mining's land for display at the annual Pageant of Peace outside the White House in Washington, DC, 1970. *Courtesy of Jim Scott*

general manager up to an overlook that surveyed the entirety of the landscape created by the largest taconite mine at the time and told his boss, "We've got to be doing more." Despite the costs to the company, the boss agreed, and that is when Dickinson was able to begin employing both his forester's knowledge and his sportsman's passion toward the pursuit of ecological reclamation. He was a significant mover and shaker, and according to later coworker Dan Jordan, "He wanted geese on the tailings basin, deer habitat, fish in the waters. He wanted to start doing something with the mined lands."

Youngman recalls, "Pickands Mather was an open-minded company in that way. They were very open to restoring the land as close as possible to what it has been before. Just about anything that we [foresters] wanted to do, they would approve." In other words, he and Dickinson were given carte blanche to apply the best techniques they could muster into the reclamation of Erie's mine-disturbed lands.

Youngman recalls that Pickands Mather & Company decided "on their own volition" to try to reclaim the fine, inorganic taconite tailings. When the taconite process was still being developed, Erie had hired a soil scientist from the University of Minnesota, C. O. Rost. According to Youngman, "The soil scientist was retained to see if he could grow grasses and legumes on taconite tailings. Everyone thought it was so barren of nutrients that you would need to put topsoil on it. He proved that by adding N and P [nitrogen and phosphorus], that you could grow vegetation directly on taconite tailings." When Dickinson learned of these findings, he immediately began a series of test plots on the taconite tailings basin to examine various types of vegetation and fertilizer. Emphasizing the difficulty of these experiments, Youngman recalls that the tailings basin was "like a low desert. It was just like sterile sand." In his 1971 talk, Dickinson revealed that the southern slopes of the tailings basin—because of the dark color of the coarse tailings used in dam construction—"recorded surface temperatures of 145 degrees Fahrenheit, which are due to the absorption of the sun's rays." Desertlike, indeed.

Dickinson and Youngman experimented with over fifty varieties of species from "low desert" regions, ranging from coniferous trees to garden vegetables. Because of the lack of nitrogen in the tailings, they placed special emphasis on species that would fix nitrogen from the air and thus make it available to plant life, such as alfalfa, bird's-foot trefoil, Siberian peashrub, European alder, crown and milk vetch, and sweet, alsike, and red clover. They also tested various seeding, mulching, and fertilizing techniques—and had to invent new methods to perform these routine agricultural tasks on the steep slopes of the tailings dam.

One day after years of growing vegetation on large areas of the tailings basin, Youngman dug into the ground with a spade. He observed a thin layer of black dirt, which had been created by the composted remains from earlier years of seasonal grasses and legumes. This dirt, called the "A horizon" by soil scientists, lay atop the inert desert of sandy tailings. Youngman remembers smiling and saying to himself, "It's building soil now." Dickinson echoed this sentiment when he wrote, "Probably our greatest satisfaction comes from the knowledge that we are building soil from crushed rock."[23]

Subsequent studies have found that Erie's vegetated tailings basin provides a prairie-like habitat that attracts birds not normally associated with northeast Minnesota. For example, as some of their native habitat was shrinking, sharp-tailed grouse were finding a home on the revegetated tailings basins and stockpiles. The result of Dickinson's work was recalled by Erie's operation general foreman Adrian "Ace" Barker, who said the tailings basin "was just like a park. Erie was kind of a showplace."[24]

In another example of multiple resource management, Youngman contacted a former professor and ruffed grouse expert to help develop a management area for the species, which inhabits forested spaces, especially those with scattered clearings and dense undergrowth. The pair designed a forest management plan that created a habitat for the grouse. When the area was ready, Youngman recalls that Erie "made a trail through there so people could hunt adjacent to the three-mile road between the plant and Hoyt Lakes." He himself walked the trail and recalls there "were plenty of birds in there."

Erie's experiments were not limited to terrestrial habitats. Youngman recalled an assignment he received from Erie general manager Clyde Keith, whom Youngman describes as "a conservationist in his own right." Keith had asked him to buy minnow traps and put them in "every water body on the mine you can think of." Youngman deployed the traps in flooded taconite pits, reclaimed cells of the tailings basin, and the former natural ore pits on Erie's land. He said, "I found minnows everywhere!" Keith explained he was interested in knowing if the mine pits could be used for recreational fishing in the future. Having seen evidence that Erie's pits could sustain fish, he began thinking about reclamation activities that could create underwater habitat for fish in the pits after mining was completed. Youngman recalls a hydrology study performed on one of the pits to determine the long-term water levels that would result after mine dewatering ceased. Once the elevation was known, Youngman recalls, "We planted that whole area with grasses and legumes"—to build an organic base for the future pit lake's littoral zone, or marshy perimeter. "We seeded it in with willows for the shoreline. Then the water covered it. Now there is habitat down there for minnows and fish. I am not a betting man, but I would bet that most of those taconite pits are filled with good-sized fish."

Dickinson and Keith were not the only Pickands Mather employees who promoted how the company's vision of multiple resource management could have a positive impact on the future use of minelands. In a

1971 presentation, Thomas Manthey, director of public affairs at Pick-ands Mather headquarters in Cleveland, Ohio, asked, "Why shouldn't the picturesque man-made canyons be advertised as tourist attractions? Why shouldn't the water reservoirs, and possibly inactive pits be developed for sport and recreational use? Why shouldn't stockpiles be developed to enhance the landscape and to permit future public use?" Manthey concluded that the potential of mineland reclamation and repurposing was "limited only by man's creative imagination."[25]

In 1977, Dickinson's visionary efforts in voluntary mineland reclamation achieved national recognition. Erie Mining Company received the first National Environmental Industry Award for Excellence, jointly awarded by the relatively new US Environmental Protection Agency, the President's Council on Environmental Quality, and the Environmental Industry Council. Dickinson and Pickands Mather president Elton Hoyt III flew to Washington, DC, to receive the award. The value of this recognition is sharpened when compared with the criticism and heavy scrutiny that was being leveled at other sectors of the mining industry at the time. That same year, the US Congress passed the Surface Mining Control and Reclamation Act (SMCRA), which regulated the coal mining industry. Federal officials lambasted that industry for "destroying or diminishing the utility of land for commercial, industrial, residential, recreational, agricultural, and forestry purposes by causing erosion and landslides, by contributing to floods, by polluting the water, by destroying fish and wildlife habitats, by impairing natural beauty, by damaging property and citizens, by creating hazards dangerous to life and property, by degrading the quality of life in local communities and by counteracting governmental programs and efforts to conserve soil, water and other natural resources."[26]

In contrast, Erie's voluntary philosophy of multiple resource management sought to enhance the quality of life in the local community by promoting multiple uses of company lands for industry, recreation, wildlife, and forestry. It was an act of courageous forethought. Youngman, in a 1992 paper entitled "Mineland Reclamation at LTV Steel Mining Company," wrote, "Under the able leadership of Sam Dickinson, research knowledge was refined and applied to large scale reclamation plantings. By the late 1970s, we had tested hundreds of plant species and reclaimed thousands of acres of stockpiles, pitwall and

In 1977, Erie Mining Company received the first National Environmental Industry Award for Excellence, jointly awarded by the relatively new US Environmental Protection Agency, the President's Council on Environmental Quality, and the Environmental Industry Council. *St. Louis County Historical Society*

tailing basin areas." He noted that visionaries like Dickinson had "laid the foundation of an idea that would one day turn thousands of acres of mining waste on the Mesabi Iron Range back into productive land."[27]

Dickinson retired from Erie Mining Company one year after Pickands Mather received the national award. His work—and that of other mine reclamation pioneers, such as Gilbert Leisman and John Dougherty—helped to prove it was possible to reclaim iron mining lands though the unintentional acts of nature and the purposeful actions of people. Of course, not every reclamation experiment proved to be a long-term success. Some of the experimental plantings employed nonnative species, such as bird's-foot trefoil and black locust, that—while extremely useful for quickly stabilizing disturbed soil or creating a thick wall of thorny "living fence" that would keep people away from dangerous mine areas—proved in the long run to overwhelm the native plants and eventually led to areas dominated by an ecologically stunted monoculture. Of such challenges, Dickinson wrote that "there have been frustrations and failures" and "the results of our work in revegetation have not been an overnight success by any means." Still, he and the other young reclamationists proved that mined land was not wasteland and could provide unexpected value as wildlife habitat, recreational areas, and forestry resources. Reflecting upon this time of experimentation over forty years later, former Minnesota Department of Natural Resources mineland reclamation supervisor Julie Jordan recalled, "Some of the trees that Sam and Dave planted—they are just fabulous reclamation. At first, you could tell what they planted, but now it's just part of the regular landscape up there. It is beautiful what they did—with the stockpiles in particular." Dan Jordan, former Iron Range Resources and Rehabilitation Board mineland reclamation director and spouse to Julie, added that "Sam and Dave were working way beyond their years in figuring out mine reclamation."[28]

The wave of environmental regulation that characterized the 1970s and early 1980s would eventually catch up to the Minnesota mining industries, resulting in the end of strictly voluntary, experimental reclamation. Around the same time, a multiyear mining bust would bring the taconite industry and its dependent communities to their knees. These forces were to have an effect on the newly retired Dickinson. Under the changes brought by the mining downturn, Minnesota's 1980 Mineland Reclamation Rules, and one of the nuances of the state's taconite tax amendment, his career proved to be far from over.

· CHAPTER 4 ·

Rules and Agencies

Only one state in the union has done less in mine-land reclamation than Minnesota.

Governor Rudy Perpich, State of the State Address, 1977

My memory of the day is strange enough that I sometimes wonder whether it really happened. My feet were on the sun-warm concrete curb, and my eight-year-old arms were extended up toward the July sky. Candy sailed through the air as the marching music of the high school band filled my ears. But my arms were not outstretched to catch a Jolly Rancher or Gobstopper; they were reaching for a baseball bat. The mustachioed United Steelworkers union member was handing it around so the parade-goers in my mining hometown could take a swing at their company's float—a battered Datsun being dragged behind a truck. To the old men in town, it symbolized the good American jobs that had been stolen by foreign automakers. The baseball bat was meant to exorcise the frustrations of the unemployed iron miners and families—men (they were mostly men at that time) whose jobs had been cut when American cars were supplanted in the marketplace by foreign models. A good dent in the Datsun might smash away some the frustrations of lost income, the feelings of helplessness, and the general decline of our morale and our towns. But my brother had been laid off from the mine and now drove his Mazda back to school to be retrained. And my father was no longer a miner. He was a first-generation college graduate who—though he had worked in the mines during his late teens with his father—had made it through Michigan Tech and now was a mechanical engineer employed by the air force. And so, when my dad locked onto my eyes and curtly shook his head, I lowered my arms and the steelworker walked on. Others could swing the bat.

As an Iron Range child during a mining downturn (in my case, in Negaunee, Michigan), public outlash was not new to me. For us, the 1980s didn't mean preppies and "greed is good"; it meant we were witness to physical expressions of frustration at an economic situation that was more complex than we could understand. The hard times at home often translated into hard times at school. I vividly recall two girls who were classmates fighting each other in the hallways—tearing at black heavy-metal T-shirts and pulling on ratted-up hair. My cousin Al fought regularly, and occasionally I got dragged in. It simply seemed normal to me until I became a parent and realized that, although we still lived in an Iron Range town that was sometimes plagued by mining downturns, my children never got punched in the head.

With the benefit of age, I can appreciate the complex global pressures that affected my community during the era in which I grew up. The late 1970s and early 1980s were devastating to the taconite industry and the towns that depended on it. My father-in-law, who worked for a railroad that hauled taconite pellets from the mine to the ore docks, said he was laid off more often than he worked for a period of five years. Similar circumstances faced Erie Mining Company, which cut production from double digits to single digits in millions of pellet tons per year, which is why Sam Dickinson retired in 1978. Dickinson chose to leave Erie Mining Company so that his hand-picked protégé, Dave Youngman, could stay on.

But Dickinson was not the type to easily stay retired. By 1979, the state of Minnesota was holding hearings and tours of the Mesabi Range's mine-affected landscape to gather comments on rules that would require the reclamation of iron mining lands in Minnesota. These rules, which were officially codified as Minnesota Rule 6130, were better known as the "Taconite and Iron Ore Mineland Reclamation Rules"—or the Red Book, based on the color of the booklet that contained them. Many had been written to capture the best practices Dickinson had developed through the voluntary reclamation activities at Erie. He was so familiar with them that he felt a calling to consult to the regulators and the industry he had retired from to help them implement the new rules. Thus, only a few months after leaving Erie, Dickinson joined Barr Engineering Co. as an environmental consultant.

At Barr, his land reclamation and forestry expertise was complemented by that of engineers and other natural resource scientists.

There were geotechnical engineers who specialized in tailings dam construction and stockpile stabilization and hydrologists who could predict the water levels in mine pits years before they were stabilized so that reclamation plantings could be targeted to the right location and habitat. Phil Solseng was one such person.

Solseng remembers, "Sam did all the early [mineland reclamation] work at Barr in the late 1970s and early 1980s. He was really the founder of it, in my opinion. I just helped Sam with the engineering work—how to stabilize it, shape it, adjust for the water levels, make sure it could grow grass. Sam was the guy who understood the reclamation, the way to do it."

The new rules tackled a range of challenging subjects, from where mines could be located (relative to areas occupied by the public and important environmental resources) to how blasting with explosives could occur to how the stockpiles, pits, and tailings basins should be shaped and revegetated to promote a useful life after mining. They required mines to have a permit, and eventually they would require mining companies to set aside money as a guarantee that mandatory reclamation could be performed after mining was completed.

The rules came into effect on August 25, 1980, but they had been in the works since 1968, when the Minnesota legislature passed Statute 93. The period in between was the era of highly publicized environmental disasters followed by landmark environmental legislation across the country. The Cuyahoga River had caught fire in Ohio, and inexplicable ailments developed in residents of Love Canal in upstate New York. The US Environmental Protection Agency was created. The National Environmental Protection Act and the Clean Water Act were passed. And in 1977 the coal mining industry was regulated by the Surface Mining Control and Reclamation Act. Solseng remembers that environmental regulation "was the ambiance of the time. There were some big disasters that had occurred. Everyone thought we should have some rules for mining."[1]

One of the most prominent environmental disasters took place in 1966 in Aberfan, Wales, where a coal mine tip—the British name for a spoil or waste rock pile—slid into a public school, killing 116 children and 28 adults. (The disaster was depicted in Season 3 of the television program *The Crown*.) Investigation of the catastrophe showed that the coal tip had liquefied because it had been built over a spring and was further lubricated by 6.5 inches of rainfall over a three-week period. The combination of unstable ground conditions and excessive rain created a landslide from the coal tip, which had been left at a steep angle with no vegetation or benching to stabilize it.

The new rules sought to prevent such occurrences and were, understandably, focused on stabilizing and revegetating mined lands. After Aberfan, the United Kingdom took regulatory actions to prevent the haphazard disposal of mine waste with "the basic policy being that all tips should be treated as civil engineering earthworks." The Minnesota rules were the same. Stockpiles could not be built higher than thirty or forty feet (for rock stockpiles and overburden piles, respectively) without integrating a "catch bench," or flat terrace, before adding another thirty- or forty-foot lift. Also, overburden piles and the overburden portions of pit walls—which were composed of glacially rounded materials that could more easily roll than angular rock—had to be laid back at a slope of 2.5 feet of horizontal run to each foot of rise (a ratio of 2.5:1), as well as being interspersed with flat catch benches. The catch bench was meant to prevent large landslides as well as to serve as a platform for trees and other vegetation to easily take hold. In practice, this meant mining landforms would have to change shape: from steep-sided, amorphous blobs built more for convenience than safety to tiered, engineered structures that looked more like step pyramids.[2]

These rules were not put into effect without intense debate. The public, elected officials, civil servants, outside experts, the mining industry, and environmental groups like the Sierra Club all participated. One subset of the mining industry, known as scram miners, or "scrammers"—small firms mining the last remnants of natural ore as opposed to large firms performing taconite mining—sought total exemption from the reclamation rule. Records from the hearing show that four public meetings and one mine tour were held over the course of four weeks in October and November 1979. According to the report of the hearing examiner, the shape of mine-built features was central to the debate. The final configuration of mine-affected landforms was "by far the most important issue to the affected public," and lengthy consideration was given to the "dispute . . . over how flat the slopes ought to be for the edges of pits and stockpiles."[3]

Dozens of pages of testimony and reporting are dedicated to the issue of whether edges should be left at the angle of repose (the steepest

Stockpile shaped like a step pyramid at U.S. Steel's Minnesota Ore Operations (Minntac), Mountain Iron. *Courtesy of Vance Gellert*

angle material can hold without falling), at 2:1, or at 2.5:1 and how high the lifts between catch benches should be. While a lot of the hearing's documentation is engineering and scientific esoterica, it is noteworthy that the Sierra Club felt "piles ought to be kept no higher than surrounding topography"; that "persons living and working on the range were the most concerned with . . . minimizing the amount of land dedicated to mining activities"; and that the Minnesota Department of Natural Resources (MDNR) thought parks, woods, and open space were "likely to be the predominant land uses" after mining. Many of these objectives conflicted with one another—for example, if waste piles needed to be kept lower than surrounding topography, they would, by necessity, occupy a greater area and increase the amount of land dedicated to mining activities. It was ultimately up to the regulators to sort out these conflicts when they codified the rules.

Over ninety people attended the hearing and approximately thirty written comments were received (which constitutes a major response on the sparsely populated Iron Range, especially right before deer hunting season). Among those who testified were Sam Dickinson and a young forestry ecologist from the University of Wisconsin–Stevens

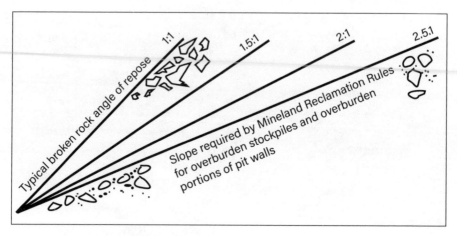

Slope angles. *Courtesy of Louise Lundin*

Point named John Borovsky. Dickinson and Borovsky spoke of the second-most hotly debated topic in the new rules: the revegetation of minelands and the ideal thickness of rooting zone material that should be placed to encourage it.

The rules required that overburden piles, lean ore and waste rock stockpiles, tailings basins, the nonrock portions of pit walls, roads, and plant areas—basically every place disturbed by mining except bare rock areas and those that would be underwater after mining—be actively revegetated. In other words, they must be planted and nurtured like a garden. The progress of vegetation would be measured at intervals of three, five, and ten years after mining in comparison to an "approved reference area." That meant it was not an absolute numerical standard, such as a requirement that drinking water contain no more than ten parts per million of nitrates, but instead a relative standard under which success is measured against an already reclaimed area. It was a new concept to many; Dickinson in his testimony described it as "a rather abstract conception," saying that, at first blush, it was "rather hard to conceive just what was meant." Barr was brought under contract to the MDNR to help identify reference areas to be used by the regulators and industry during implementation of the new rule.

The advantage of the reference area concept was that the revegetation success of individual portions of a mine was measured only against a similar piece of mineland that had been successfully revegetated. In other words, it was inherently recognized that minelands would not look like virgin forest after only ten years of reclamation. But the rule assumed it was reasonable to expect that the south-facing slope of a lean ore stockpile could attain the same vegetative success of a similar reference stockpile that had been successfully revegetated to the MDNR's satisfaction ten years earlier. Some of the reference areas that Barr selected were located on Erie Mining Company's lands, which meant that much of Dickinson's pioneering work now became the model against which reclamation would be measured elsewhere on the Iron Range. Another advantage of the reference concept is that drought or plant disease might affect both the reclamation area and the reference area equally; therefore, mining companies would not be held liable for meeting a strict numerical standard when factors outside their control were at play.

The new rules required two feet of loose material (typically stripped overburden) to be placed over the flat areas of rock stockpiles to serve as the rooting zone for vegetation. This requirement was meant to address the phenomenon Gilbert Leisman had observed nearly thirty years earlier: that bare rock stockpiles, even if they had trees, typically had too little groundcover to build soil. Julie Jordan recalls that the two-foot cover thickness was determined by a "wedge study." In this study, rooting zone material was placed in a wedge shape varying in thickness from essentially no cover (the tip of the wedge, like the cutting edge of an ax) up to five feet of cover. Various types of grass, trees, and ground-covering herbs were tested in rows planted across the span of the ax-head. It was determined that two feet of cover, on average, produced the best growth for all types of vegetation.

The required thickness of loose groundcover for vegetation to take root was (and remains) debatable and was hotly contested in the hearings. Decades earlier, Leisman had documented that some trees—such as birch, red pine, and jack pine—could grow on bare mine rock, and natural examples were easily visible on Minnesota's North Shore and in the Boundary Waters Canoe Area, where pines and other trees cling to sheer rock cliffs. In early testimony, the mining industry pointed out volunteer tree growth on rock stockpiles. But the hearing examiner's report later stated, "Industry allegations that that rock storage piles develop naturally screening vegetation was contrary to the facts as disclosed during the tour." In other words, the natural volunteer tree growth was typically too sparse and too slow to meet the new standards. Thus, it was established that zero cover was insufficient. But the wedge study had shown that the optimal cover thickness was, in fact, dependent on the *type* of vegetation that was being planted—shallow-rooted plants required less cover soil to thrive than deep-rooted plants. Therefore, given this variability, the rules allowed mines to propose cover thicknesses less than two feet, as long as they could demonstrate through acceptable research that the thinner cover could grow dense vegetation that was self-sustaining in ten years or less.

Although Erie Mining had pioneered and won awards for voluntary reclamation work, the company had to comply with the new rules along with the rest of the mining industry. Based on years of prior experimentation, it felt it could meet the new standards without incurring the costs associated with placing two feet of cover. To meet the ten-year reclamation timeline required in the rules, Erie planned to vegetate its stockpiles with grasses and legumes. To provide the rooting zone material, it would cover the stockpiles with overburden (which resembled unsorted glacial till after stripping, whether it had originally been sorted or stratified through glacial deposition processes or not), but it planned to test the optimal cover thickness through its own research. Dave Youngman recalls Erie's approach: "Originally, the DNR wanted two feet of cover on everything. We had to demonstrate that we could meet the standards with less. So we planted test plots with two feet of glacial till, one foot of glacial till, and 0.5 foot of glacial till. Grasses and legumes grew just as well on six inches of till as they did on twenty-four inches." Based on this demonstration, Erie was granted a variance allowing it to use a thinner sheath of cover material over its rock reclamation sites than the rules required.

In order to meet the vegetation timeline noted in the new standard— which required self-sustaining vegetation within ten years after mining ended—many mining companies turned to rapid-growing, nonnative groundcover species, such as smooth brome, Canada bluegrass, bird's-foot trefoil, and cicer milk vetch. Besides their quick growth, these species were drought tolerant (in case there was insufficient precipitation during the ten-year reclamation period) and had root systems that could stay in place during intense rainstorms (thereby keeping underlying materials from eroding). But these species tended to dominate a reclaimed area, creating a dense mat of vegetation that crowded out native species, including trees. In this way, it seemed the rules' rapid revegetation requirements created conditions that limited the return of native forests. Youngman said, "It *hindered* that development because it was so dense. But it didn't stop it. It came slowly. There was a natural succession, but it was a slower succession."

In Youngman's experience with voluntary reclamation prior to the rules, trees grew more rapidly without groundcover: "If we wanted trees in an area, we would plant trees and not ground cover." But tree planting alone was a gamble with respect to meeting the timeline the rule required for groundcover. Thus, another unintended consequence of the mine reclamation rules was they tended to create more grassland habitat than forest. In recent years, similar observations have been made in mountaintop coal mines, where it has been found that well-intended reclamation efforts have "left behind a landscape of angular plateaus and open prairies alien to the Appalachians." Recent work has shown that tree planting alone can be effective in restoring coal mined lands, a method known as the Forestry Reclamation Approach.[4]

Under the new rules, as noted, the upper portion of open mine pits also needed to be resloped from angle of repose to 2.5:1 with a catch bench at the toe. This rule was created to allow for revegetation of the nonrock portions of the mine pit rim, but more importantly was aimed at public safety. With over four hundred open-pit mines across the Mesabi, and with communities built right next to mines, the edges of mine pits were everyday dangers for Mesabi residents. Statistics from the era showed that one death or injury occurred every three years involving nonminers in open mine pits (see Appendix 1).[5]

The original method to deal with the hazard of falling into pits had been a state law passed in 1950 requiring all open-pit mines and shafts to be fenced off or physically blocked by equivalent means. Minnesota Statute 180 was (and still is) enforced by county mine inspectors in each county where mining was active. This law ensured a physical barrier to prevent the public from accidentally walking into a mine pit; however, it was a less-than-perfect solution that did nothing to address the steepness of the pit walls themselves. In some areas, the steep

Houses abandoned due to proximity to mining, North Hibbing, 1941. *Library of Congress*

pit walls eroded back beyond the fence line, leaving some portions of fencing hanging in the open air, pinned to the ground that remained solid. In other locations, the public simply cut the fence to get to their favorite trails, hunting areas, or swimming holes in flooded mine pits (none of which were legal). And in counties where mining had ceased, there was no authority to enforce the law. The new sloping requirements of Minnesota Rule 6130 sought to correct these deficiencies by addressing the hazard itself—the steepness of the pit walls.

A final piece of the hearing, which referred to only brief phrases in the rule, heralded some new considerations for many operators of Minnesota iron mines. Among the goals for mineland vegetation were "to screen mining areas from noncompatible uses" and "provide wildlife habitat or other uses such as pasture or timber land." In other words, vegetation should limit the public's view of mining *and* create a basis for the productive use of the land after mining. The first phrase implied that perhaps, eighty years after the steam shovel en-

A mine fence meant to prevent trespassing dangles along an eroded pit edge.
Author's collection

abled the mining industry to shape the landscape, some people were growing tired of looking at mining activities. The US Department of the Interior touched on this topic in 1967, when it studied the Iron Range for the report "Surface Mining and Our Environment," which explored the status of mining prior to regulation:

Iron-ore open pits extend over large areas and usually reach considerable depths. The Mesabi Range of Minnesota is an iron ore formation 120 miles long and 3 miles wide. In the next hundred years the Range could become a giant canal or lake. The minerals in the formation are chemically inert and the terrain is flat; thus, mining operations cause little or no water pollution. Many of the pits have steep highwalls, which, in places, are dangerous and impede the movement of wildlife and humans. The tremendous piles of low-grade, or "lean" ore that dot the countryside may either be considered an *attractive feature in an otherwise monotonous landscape* or as *unsightly, depending on one's viewpoint.* [emphasis added][6]

Whether mine-created landforms are beautiful or unsightly remains a topic of debate among residents and visitors on the Iron Range. Regardless of anyone's opinion, nothing could be done to force companies or landowners to limit the visibility of the mined landscape they created prior to 1980; thus, the new rule would shield the portion of the public that considered mining to be unsightly from seeing future taconite mining and its wastes. Seeking clarification on this topic, a representative of a now-defunct mining company asked the MDNR at the hearing whether the rule had an aesthetic requirement, and if so, whether it should be applied only near population centers where there were people who might see mine waste.

> MDNR spokesperson: It isn't an aesthetic requirement *only* . . . it is providing a productive subsequent use . . . a hidden stockpile today is likely to become visible tomorrow . . . more and more people are taking to the woods and looking at these things. [emphasis added]

> Mining company representative: I have a little problem with that, just a little one. That just indicates that we are going to have the greatest planting operation undertaken since Weyerhaeuser discovered you could manage forests and I'm wondering if it will be absolutely essential in all cases for that five-backpacker-per-month or -year to plant the benches of rock strip piles [lean ore and/or waste rock piles] that are five, ten miles back in the woods.[7]

In other words, was it worth millions of dollars to prevent some backpackers who happen to wander near a stockpile miles away from town from seeing any barren land? In the years since Theodore Roosevelt's time, had the country moved beyond "not one cent for scenery"?

> MDNR spokesperson: It is absolutely essential a subsequent use
> be provided for[,] and to me, in this part of the country, the most
> logical subsequent use for much of this land is timber.

John Borovsky, a graduate student from the University of Wisconsin and later my colleague at Barr, added, "We would like to think in natural resource management we are capable of imagining more than one use. . . . The words I'm talking about are what foresters refer to as Multiple Resource Management, Multiple Use Management. . . . [T]here are other considerations besides the timber, the commodities that are produced there." He ended by saying, "Where you have a visual resource to work with, it is your responsibility to manage that visual resource and to minimize the impacts on that resource while producing commodities from the area."

After the questioning, the phrase was kept in the final rule. And just a few years later, the very mining company whose representative had asked the pointed question retained Borovsky as a consultant, indicating acceptance of and intent to comply with the new standard by the portion of the mining industry that was unfamiliar with these new concepts. Although the phrases that sparked the debate were small, their retention in the final rule indicated that a little bit of Sam Dickinson's and Erie Mining Company's visionary way of thinking about mineland management and reclamation—that the land had multiple uses and should be managed to provide the greatest benefit to the greatest number of people—had made it into the state rules.

■ ■ ■

In the process of finalizing the mine reclamation rules, the question arose about what to do about all the mine-disturbed land that had been created in the ninety years *prior* to regulation. As described in previous chapters, the development of mechanized mining, natural ore washing, and the first twenty-three years of taconite beneficiation had left a massive footprint upon the land. By 1980, iron mining had disturbed somewhere between eighty thousand and eighty-five thousand acres of Mesabi land. And as noted during the hearings, the erosive potential, nonvegetated nature, and visual impact of the minescape were becoming public concerns. The proposed solution to this conundrum was twofold: 1) incentivize scram mining companies to finish mining the remnants of natural ore, then reclaim those areas to modern standards, and 2) create a state governmental division tasked with reclaiming the pre-1980 minelands.

As noted, the scram miners—as smaller firms with arguably fewer resources to dedicate to reclamation—originally wanted to be left out of the Red Book. While their request was denied, scram mining was incentivized in other ways. These companies were exempt from certain formal environmental review requirements that applied to the taconite industries. Small scram mines (less than eighty acres in size and lasting less than five years) were exempt from some aspects of permitting. But miners of residual natural iron ore resources were still required to perform reclamation. The philosophy was like that of a china shop: if you broke it, you bought it. If you disturbed a pre-1980 stockpile, open pit, or tailings basin for the purpose of scram mining, you would need to reclaim it to the new standard. In this way, it was thought that scram miners would chip away at the ninety years of stockpiled lean ore people such as Greenway had set aside, thus reducing the volume and visibility of the disturbed landscape and slowly bringing it into compliance with the reclamation standard.[8]

By 1979, the second part of this strategy was fully implemented when a Mineland Reclamation Division was created at the Iron Range Resources and Rehabilitation (IRRR) Board. The IRRR originally opened for business on July 1, 1941, and in its first decades initiated projects in such fields as county land use, geological and hydrological studies, tree planting, aerial forest surveys, and diversified economic development. During World War II, the agency's funding of studies related to powdered iron, sponge iron, and taconite anticipated the postwar technological transformation of iron mining in the Lake Superior district—for example, it helped fund E. W. Davis and his taconite work with the Mines Experiment Station at the University of Minnesota. In short, the role of the IRRR was "to develop jobs and income in those counties affected by the decreasing amounts of taxes created by iron ore" that occurred as iron production slowed and as companies transitioned from natural ore mining to taconite. An example of its investment in economically diversifying the Iron Range—to better enable the region to weather both the decrease in tax revenue and the bust cycles of mining—was start-up funding for Jeno Paulucci's agriculture and canning business for Chinese food, a venture that eventually grew into the $60 million Chun King canned food dynasty. Other industries the IRRR invested in were peat mining, forestry, light manufacturing, and, importantly, tourism.[9]

The new IRRR Mineland Reclamation Division was tasked with reclaiming state-owned lands that had been mined prior to 1980. While these were exempt from the Red Book, based on both the period in which they were mined and the fact that the state of Minnesota did not operate any mines, their reclamation would accomplish several of the IRRR's objectives: 1) improving safety by eliminating dangerous areas, 2) establishing vegetation and reforestation, 3) repairing and preventing erosion and dust problems, 4) creating recreational areas, and 5) restoring wildlife habitats on abandoned minelands. By implementing projects based on these goals around the time that the 1980 rules were being put into effect, the IRRR would also create "demonstration projects"—on-the-ground examples highlighting methods that could be employed to reclaim mined lands in Minnesota.

Implementing mineland reclamation techniques on a ninety-year-old mining range that had supplied three-quarters of the iron for the development of the nation's infrastructure proved to be no small feat. Phil Solseng said of the scale of the challenge faced by the Mineland Reclamation Division, "Minnesota's Iron Range was the biggest reclamation project in the world. And the IRRR really led that." And Dickinson—in his role as a reclamation consultant at Barr with a team of complementary engineers and scientists—was hired by the IRRR to help make it happen. Solseng recalled, "Barr did most of the early work for IRRR. We tried to set the standard of how things should be done. The St. James Mine was key as a showcase."

． ． ．

The St. James is a former open-pit natural iron ore mine bordering Aurora, Minnesota. By 1980, groundwater had reentered the pit after mining operations had ceased and the city of Aurora was using it to supply water to the townspeople. While the groundwater entering the pit proved a source of abundant, nearby, and high-quality drinking water, the reddish clays and ore fragments in the pit walls were unstable and would occasionally erode into the pit waters. Solseng remembered, "It would turn red every spring from runoff." He was told that if there was a rainstorm or snowmelt event on the weekend, "on Monday morning, everyone's white shirts were pink."

Solseng, Dickinson, and others at Barr designed a plan to shape and stabilize the eroding pit walls according to the new rules. They also filled the shaft of an underground mine to prevent people from accidentally stumbling into it. They covered a rock stockpile with overburden and planted it with pine seedlings. Consistent with Dickinson's philosophy of multiple resource management, they designed a public park on the east side of the pit. Sand was dumped on the bottom of the shallower parts of the pit so that fish would have littoral spawning habitat instead of just a sterile four-hundred-foot-deep bathtub to swim around in.

Upon completion, the pit reopened as the public water supply, a fishing destination, and a public park. Eventually, a paved trail was added around the pit to create opportunities for walking, running, bicycling, and other nonmotorized uses. The only challenge that remained unsolved was the final stabilization of water levels in the pit. Because Erie was still mining nearby and Aurora was pumping the pit for its supply, the region's overall water table had not yet recovered to its final elevation; therefore, that aspect of the design would need to wait for groundwater levels to stabilize, at which time the final outfall could be constructed. (The water levels are still recovering today.)

The St. James Mine proved to be, in Solseng's words, a "poster child for good reclamation." In 1981, it won the Seven Wonders of Engineering Award. And, importantly, Solseng remembers that "people liked it."

Barr went on to design similar "showcase" projects for the IRRR at the Judson Pit near Buhl and the Carlz Pit near Keewatin. Each of these involved a lot of earthwork to shape and stabilize the mine-created landforms and then revegetate them. Visitors to the sites will notice a prevalence of red pine and jack pine trees, which were favorites of Sam Dickinson and well suited to the unique demands of mine

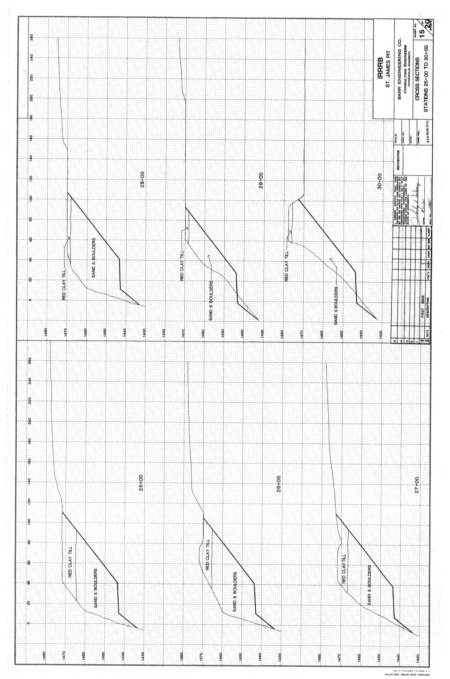

The reclamation design for St. James pit included plans to shape and stabilize the pit walls. *Barr Engineering Co.*

The St. James pit now invites recreational activities. *Author's collection*

pit reclamation. Today, these pines tower forty or fifty feet overhead, susurrating with the wind and coating the ground with pine-needle litter that builds the soil.

In the early 1980s, Dickinson and Solseng thought Barr's work creating showcase projects for the IRRR would lead to a long-term practice of designing reclamation projects across the Mesabi. However, state lands requiring reclamation efforts were limited, and the devastating taconite bust of my youth was in full swing. All the taconite companies, after two or more decades of growth and stability, had to shutter or slow production just to survive. One company, Butler Taconite in Nashwauk, never recovered. Butler closed its doors in 1985 and became the first (and still the only) taconite mining operation to undergo full reclamation and permit release under the new rules—a process consisting of, among other things, demolishing the plant, breaching and diverting water around the tailings basin so it no longer held water, and shaping and revegetating the disturbed lands. The process took about eight years and cost roughly $3,368,000. While most taconite operations weathered the downturn in one way or another (including closing and then reopening under new ownership), there was no money to consult with outside reclamation engineers and scientists. It was becoming clear to Dickinson that if he wanted to continue playing a part in the rollout of mandatory mineland reclamation practices in Minnesota, he would need another job. Therefore, by the mid-1980s, he left Barr for public service with his largest client, the IRRR Mineland Reclamation Division.[10]

Dan Jordan remembers meeting Dickinson at the IRRR: "He knew a lot—from the school of hard knocks mostly—and he was an excellent sharer of his knowledge." Dickinson started at the IRRR as he had started at Erie—more as a forester than a reclamationist. But in relatively short order he was reengaged with the mine reclamation work he loved. The newly formed Division of Mineland Reclamation was, at that time, located in the state-owned abandoned buildings and mine workings of the former Hill Annex natural ore open-pit mine near Calumet, Minnesota. It proved to be the perfect site for Dickinson and Jordan to demonstrate rule-compliant mine reclamation practices as well as test new concepts. Jordan recalled, "We'd come up with these ideas that utilized our collective knowledge and challenged our expertise. Then we would take them to the 'lab' for testing. Remember, our 'testing lab' was 640 acres, one

View of the Hill Annex Mine, 1970, which later became a testing ground for mine reclamation techniques. *Minnesota Discovery Center*

square mile, on the north side of the Hill Annex Mine to use as a proving ground before taking them to the field."

The pair tagged and measured the trees that were being used for reclamation, watching them over years to see how they thrived on their test sites. At the time, the IRRR was buying tree seedlings from the open market. According to state law, the tree contracts were open for public bidding, with the contract going to the lowest bidder. But Dickinson's and Jordan's measurements showed that this process was not producing the quality of tree stock needed to meet the revegetation requirements of the new rules. It came to a head just a few years into their experimentation. "In the third year of the public bidding process, the quality of the containerized tree seedlings we received were so poor we did not plant about 60 percent of them," Jordan said. "They would not have survived the conditions we would be planting them in. And so the commissioner [of the IRRR] said, 'Let's grow our own.'"

At the Hill Annex Mine, a building that had once been the miners' clubhouse sat on the west side of the pit. The first floor had a dining room and social activity center. The second floor had sleeping rooms for single miners. But the hidden gem proved to be the basement, where the Hill Annex miners had tucked away a one-lane bowling alley. It had just the space and controlled environment Dickinson and Jordan needed to fulfill the commissioner's request. Jordan said, "That one-lane bowling alley became our growth chamber." With that, the IRRR rapidly and frugally entered the business of growing trees.

Dickinson and Jordan researched ways to grow better trees than those they had gotten from the open market. At the time, the British Columbia forest service had a new technique in which trees were grown in "styroblocks," polystyrene container blocks with cavities that would be filled with soil and then planted with tree seeds. Once the seedlings had reached the size needed for planting, they were easily transported to the mine reclamation site, where they would be pulled from the styroblock. The plug containing both the soil and the seedling could then be planted in a small, shallow hole rapidly created by an iron planting bar. The styroblocks would be sterilized and reused for the next crop of trees.

A former bowling alley at Hill Annex Mine became a growth chamber for seedlings destined to reforest previously mined landscapes. Millions of trees from the Iron Range Resources and Rehabilitation (IRRR) growth chambers were eventually planted across the Iron Range. *Minnesota Discovery Center*

Pine seedlings growing in styroblock, Hill Annex Mine. *Minnesota Discovery Center*

Using this same technique, the IRRR began producing forests full of seedlings. Jordan recalls, "One hundred sixty thousand trees per year came out of the old miners' bowling alley." With a four- to five-month germination period, the IRRR could grow two crops per year under metal-halide lamps, which produced the light and heat the seedlings needed to grow. In this way, trees were grown year-round and were made ready for both spring and fall plantings.

By the 1980s, "industrial tourism" was beginning to play a role in the Iron Range economy. The MDNR was conducting public tours of the Soudan underground mine near Tower, on the Vermilion Range, and Governor Rudy Perpich, the only born-and-bred Iron Ranger to ever hold the office, felt that a companion open-pit mine should also be made available for public tours. Therefore, the Hill Annex was converted into a state park which eventually was operated by the MDNR, and the Division of Mineland Reclamation was relocated to a new office in Chisholm. Dickinson retired for a second time, leaving the IRRR in 1988.

Bus tours sponsored by Iron Range Resources and Rehabilitation (IRRR) brought visitors to Hill Annex State Park to experience the inside of a vast open-pit mine, 1979. *Minnesota Discovery Center*

Dan Jordan remembers the time spent at the IRRR with "the father of mineland reclamation" as one of bootstrapping experimentation: "We didn't have big budgets for research, nor did we have researchers for the sake of researchers; we had doers. Come up with a sound idea, and we'll do it." Dickinson's and Jordan's experimental work had a lasting effect on the Iron Range and IRRR. By the 1990s the IRRR was planting three hundred thousand trees per year. There are now millions of IRRR-planted trees populating the minelands of the Mesabi, Vermilion, and Cuyuna Iron Ranges.

During this era, the IRRR became known for taking reclamation to the point that the public could reuse the minelands. The Red Book only required mines to leave mined land in a condition that would not *preclude* a future end use. But Dickinson and Jordan proved that minelands could be reclaimed to the new standards *and* have a safe and productive future use, embodying the principles of multiple resource management. Examples of IRRR projects conducted at the time include the construction of public access for trout fishing in stocked pit lakes, sledding hills on steep-sided rock stockpiles, and public swimming beaches with playful names like Lake Ore-be-gone in Gilbert.

Jordan remembers that the IRRR mineland reclamation staff was inspired by the potential for repurposing mined lands. When it came to new ideas, they thought "there wasn't anything bad out there. There were only ways to enrich this [minescape]," he said. Jordan felt the minescape left from the natural ore and wash ore operations could be embraced by the communities in which it was embedded. He noted that "the dumps were getting swallowed up by the communities" and that reclaiming them had a dual purpose of both conserving the residual iron units, should the lean ore ever prove economical to mine, and using them for other purposes: "They were there, and they were there to be used." The IRRR saw its role as "build it, make it better, and [the dumps] could be used more."

The Division of Mineland Reclamation invested in recreation and habitat creation on the minescape. A 1993 report stated that the agency was doing "about 20 to 30 mine reclamation projects annually, including mine pit accesses [for fishing and swimming], campgrounds, sliding hills, habitat for sharptail grouse, and Peregrine falcon release areas/habitat."[11]

After Dickinson's departure, Dan Jordan carried on with the division for nearly three more decades. Meanwhile, Dickinson's second retirement proved as short-lived as his first, and he was soon consulting for the state of Oregon on the reclamation of aggregate mining operations there. Although he was now past the age when most scientists stayed indoors, he spent two summers performing fieldwork to share his knowledge on how to reclaim the mined lands. Dickinson also worked on the reclamation of minelands in Sudbury, Ontario.

During this time, Dickinson continued to collaborate with his former partners. He and Jordan felt that the twelve-hour PBS documentary on the history of the Iron Range, *Iron Country*, which premiered in 2000 and was hosted by noted local historian Marvin Lamppa, would have benefited from a chapter on mineland reclamation. A sequel never came to fruition, but Dickinson continued to tell the reclamation story in other ways. He reconnected with his former Barr colleagues when he cowrote the reclamation chapter for the 2015 book *Responsible Mining*, edited by Barr environmental engineer Michelle Jarvie-Eggart.

Sam Dickinson and Dave Youngman assessing vegetation at Erie Mining Company. Their pioneering reclamation work demonstrated it was possible to directly grow vegetation and build soil on sterile taconite tailings basins. *Courtesy of Kent Dickinson*

Dickinson never stopped working on his first professional love: the reclamation of taconite tailings in Minnesota. Julie Jordan remembers him showing up for fieldwork in Keewatin one day when in his mid-eighties, fearlessly driving his family sedan up the steep outer slope of the tailings basin dams where most people were wary of driving with a mine truck. "He was nuts," she said of his driving choices, but "he had a passion for the field and still felt he could make a difference."

Unfortunately, this project proved to be his very last job. Dan Jordan, whose professional affiliation with Sam had blossomed into friendship, remembers that "he was out in the field when he shouldn't have been, health-wise." Dan once had to take him for medical attention after a field outing. Although the Jordans knew Dickinson to be very private about his health problems, they felt they should tell his friends and professional associates that his overexertion had landed him in the hospital. When they returned to visit, his hospital room had been decorated in a way that was fitting for a man who had planted hundreds of thousands of trees: "He had been given bouquets of evergreens," Julie remembered. "There were spruce, pine, and balsam. That was the kind of friendships that Sam had."

Sam passed away in 2014 at the age of eighty-nine. Many of his coworkers from Erie, Barr, and IRRR attended his funeral in Grand Rapids. They remembered him as a fair person who did not want to take credit for work he did not do himself. They remembered his love for the outdoors, youth hockey, and maple syruping. But above all, they remembered his innovator's passion for bringing new life to the mined lands of Minnesota's Iron Ranges that played out over the industry's transition from experimentation to mandatory action. His devotion to the field and instinct to share his knowledge were evident into the last days of his life, when he presented to his children a scrapbook containing photos, magazine clippings, and other memorabilia from his decades of professional experiments and accomplishments. His obituary stated that "until his death, he was still consulting for various companies that need reclamation." It was his avocation. In an understated way, typical of Dickinson, the obituary sums up his sixty years of innovation by simply describing him as "a pioneer of mine land reclamation."[12]

· CHAPTER 5 ·

Laurentian Vision

Pits and Piles into Lakes and Landscapes
Laurentian Vision Partnership tagline, 2001

Jim Swearingen had a financial problem. It was the mid-1990s, and Swearingen had recently been promoted to the role of general manager of the largest taconite mine in North America: U.S. Steel's Minnesota Ore Operations (Minntac). It was a position equivalent to that occupied ninety years earlier by John C. Greenway—namely, chief executive of U.S. Steel's flagship iron ore mining operation in Minnesota. Although Minntac was emerging from the economic woes that had plagued taconite mining and the steel industry in the 1980s (a decade in which U.S. Steel profits were nonexistent), cash was still hard to come by. Taconite mining is a business of thin profit margins during the best of times, and the economic recovery of the steel industry remained lukewarm after the ice bath of the previous decade. So Swearingen was hunting for a good idea.[1]

What he needed was a financial asset—something of value—that he could use as leverage for capital improvements to Minntac. By the 1990s, taconite mines and processing plants were beginning to show their age after being run for thirty or more years. Much of their original infrastructure was in need of replacement. To address this problem, the Iron Range Resources and Rehabilitation (IRRR) Board created a grant program, the Taconite Economic Development Fund (TEDF), which rebated a portion of the taconite production tax (levied on the mining companies per ton of taconite pellets produced) for capital improvement projects that would upgrade the aging plants. But to apply for the program, Swearingen needed to show that U.S. Steel had an asset that matched the value of the grant application—and the cash just wasn't there yet.[2]

But Swearingen was astute in these kinds of financial problems. Having worked for over seventeen years in Minntac's accounting department, he understood that the positive side of a balance sheet included far more than the cash you had in the bank. He began to mull over other kinds of assets, beyond just cash and mineral value, that Minntac could develop and count on its books. And then a casual conversation began to shed light on a potential solution to the problem. When asked who had helped him arrive at his breakthrough idea, Swearingen recalled, "Of all people—it was Tom Rukavina."

Anyone familiar with Iron Range politics will recognize the name Tom Rukavina. The *Duluth News Tribune* described the state representative for the eastern Mesabi district from 1987 to 2013 as "smart, witty, energetic, larger-than-life." Rukavina famously embodied and fought for classic Iron Range issues—from miners' rights to expanding access to alcohol at sporting events. But his career also showed an appreciation for issues for which he is not widely known, such as the value of a high-quality environment (he was an organic gardener) and nonmotorized outdoor recreation (he was a cross-country skier and former assistant director of the Giants Ridge ski area).[3]

As part of fulfilling their respective leadership roles, Swearingen and Rukavina kept up a professional relationship. But Swearingen remembers one particular conversation in which the representative prompted him to consider an issue outside the range of normal, day-to-day mining and politics: the life of Minntac after mining was completed. Rukavina was no stranger to mining landscapes, having worked three years at Minntac himself before entering political service. With his appreciation for outdoor recreation and the environment, he could see the potential of the unique landscape of mountains and pit lakes that mining had created. Swearingen recalled Rukavina painting a mental picture: "Geez, Jim, you could create ski hills! You could create lake properties!"

The seemingly offhand remark "tweaked my interest," Swearingen said. He began to contemplate the possible real estate value of the mined lands under Minntac's control—and whether it could be counted as the matching asset he needed as leverage for the TEDF grant program. After all, in the wake of the 1980s steel downturn, U.S. Steel had begun to generate cash in other parts of the country by selling off some of its real estate associated with unused or idle manufacturing lands.

In Alabama, U.S. Steel lands were being planned and developed for housing, schools, commercial buildings, and golf courses. Swearingen wondered whether a similar model could be employed on the Mesabi.

He recruited his colleague Dennis Hendricks to help study the possible value of Minntac's reclaimed minelands. Hendricks, a mine engineer, brought a wealth of subject knowledge, having worked in both mine operations and mine engineering where he oversaw some of Minntac's progressive reclamation activities to comply with the 1980 standards. At the time of Swearingen's inquiry, Hendricks had left Minntac's mine engineering department and now worked for U.S. Steel's division of resource management, which administered the company's land assets in Minnesota and other states. Hendricks understood both the technical side of mineland reclamation and real estate values.[4]

The first site where the men tested the idea for real estate development was one of Minntac's overburden dumps, located in the mine's East Pit but off the subcrop of the iron formation. It was already reclaimed in compliance with the 1980 standards, and, according to Hendricks, "it had a nice view down into the valley towards Virginia and Mountain Iron." He and Swearingen had in mind a housing development, similar to Pill Hill in Hibbing. But unlike Pill Hill, which had been built on a pre-1980 natural ore overburden and rock stockpile, it would be the first housing development on taconite lands that had been reclaimed through the mineland rules. They put together some sketches of their proposed housing development and brought them to the IRRR to serve as the basis for a potential match to their grant application. Hendricks remembered, "They didn't accept it. But they didn't completely reject the idea, either."[5]

The IRRR commissioners, while concluding that the project would not be eligible to serve as a matching asset for the tax rebate, told Hendricks it was "a pretty good idea" and left Swearingen and Hendricks with the question, "What else can you do with that?"—referring to the notion of developing reclaimed minelands for nonmining uses. It served as a form of encouragement that Swearingen and Hendricks might be onto something.

Around the time of their presentation to the IRRR, the Minnesota Department of Natural Resources (MDNR) began studying a proposal that would require mining companies to provide financial assurance for the reclamation of mined lands. Elsewhere in the country, it was becoming clear that minelands could create huge legacy costs. Under the MDNR's proposal, mining companies would be required to set aside millions of dollars for future reclamation work in the event that an unfavorable market forced premature closure of mine operations—as had been the case for Butler Taconite. Thus, mining companies would not only be obligated to perform reclamation after mining was complete (a future cost to be borne after mining and ore processing ceased), but they might also be required to pay up front for financial instruments (such as surety bonds or irrevocable letters of credit) guaranteeing that reclamation could be performed in the event of an unexpected closure. If these rules were put into effect, they would create an immediate balance sheet liability for the mining companies at a time when they were struggling to return to financial health.

And so, with a rising real estate market for the sale of former industrial lands, a potential financial incentive to rapidly reclaim mined lands, and a little official encouragement to think outside the box, Swearingen and Hendricks began to imagine that Minntac—through its earthmoving operations of 90 million tons per year—could create something of lasting value beyond the 16.3 million tons of iron ore pellets it shipped to the steel mills. Swearingen began to ask, "How can we bring Minntac to a point where—when ultimate closure comes—instead of having legacy *costs*, we have created legacy *assets*? In other words, what could we leave behind that would have *value* [for development into communities, resorts, or some other land use]?"

With this question, Swearingen and Hendricks gave voice to a movement that would eventually engage students and landscape architects from three different universities, mining engineers from all of the major taconite companies, state and local elected officials and civil servants, as well as the general populace of the Iron Range—at times in all-day and all-night work sessions—in a master visioning process about the post-mining uses of mined lands.

■ ■ ■

The project started with a phone call to U.S. Steel's real estate office in Birmingham, Alabama. In Birmingham and the surrounding area, U.S. Steel had begun to develop and sell its real estate holdings following the steel industry downturn in the mid- to late 1980s. Hendricks's counterpart, Tom Howard, was U.S. Steel's real estate general manager responsible for land sales in the region. In a 2011 interview, Howard recalled:

> Up until the 1980s, most of our land was off limits to development. We were a steel company and almost all of our land was acquired for the timber, the coal, the iron ore or to provide buffers around some of our operating facilities like Fairfield Works. . . .
> As we no longer needed the land for the purposes we acquired it for, we got more liberal in developing the land.

Thus, at the time of Hendricks's call, Howard had already been working in real estate development on postindustrial properties for a decade. His first suggestion to Hendricks was to bring in a landscape architecture professor from Auburn University named Darrell Meyer and ask him to provide a high-level opinion on whether or not the project of converting taconite minelands into legacy assets was even possible.[6]

This was the 1990s, and a lot of long-standing conventions were being broken. The Berlin Wall had just fallen, and the Cold War had ended as many former Soviet Bloc nations democratized. Hedonistic and glammy hair metal was giving way to the relatively more humane, flannel-clad grunge culture. The internet was revolutionizing everything from communications to advertising to research to sales. In this context of new ideas smashing old conventions, maybe it was not surprising that Meyer was optimistic about the project. The Alabama professor recommended that Swearingen and Hendricks work with a landscape architecture school more familiar with the vegetation, waters, and landforms of the Middle West—specifically, the University of Minnesota.

The first meeting between the landscape architects of the University of Minnesota and U.S. Steel occurred in the Twin Cities on May 7, 1998. Swearingen, Hendricks, and other key managers represented U.S. Steel's land, environment, and mine engineering departments. The university attendees included Roger Martin, dean of the College of Design, and several landscape architects, including John Koepke. At the time, Koepke was an associate professor and head of the department of landscape architecture. He was known for creating landscape designs that spotlighted the connection between culture and ecology, including work for local Native American communities (Koepke is an enrolled member of the Lac Courte Oreilles Band of Lake Superior Chippewa). This was to be the first of many meetings between Hendricks, Swearingen, and Koepke. U.S. Steel and the university found common interest in a project that would focus on the future of the Mesabi lands after mining and, according to historian James Stolpestad, "both the U.S. Steel and the University people came away from the meeting thinking they had discussed something important." With this promising introduction, a visit to the Mesabi was arranged for later that July.[7]

Hendricks recalled that first site visit with Koepke and the other landscape architects from the U of M. They were standing on a rock stockpile near Minntac's West Pit. At the time, Minntac's stockpiles would have occupied an area of more than eight hundred city blocks—larger than the entire University of Minnesota campus. They towered over the treetops, creating roughly five hundred feet of topographic relief against the backdrop of the open mine pit. The stockpiles faced south toward Mountain Iron, a town built on the edge of the mine pit, where the seven Merritt brothers had first mined Mesabi iron ore 105 years earlier. They were at the head of the major watersheds of North America, near the Laurentian Divide, from which water splits both southward (to Lake Superior and the St. Lawrence Seaway) and northward (through Canada to Hudson Bay). And it was summer on the Iron Range, a time when the verdant leaves and paper-white trunks of aspen and birch are beautifully offset by the sanguine bluffs of the bare iron formation. As they marveled at the site, Hendricks recalled one of the landscape architects shaking his head and saying, "What a challenge! It is incredibly exciting, but it also scares the hell out of me. It is so big!"[8]

In a later publication, the landscape architects would describe the mine views as "almost theatrical" in their scale and vividness. Recalling these early exposures, they described their impressions of the Mesabi as they traveled along US Highway 169: "Driving from west to east, one sees deep canyons and steep, often barren hills, defining the viewshed and creating a sublime, industrial scene made all the more vivid by the orange-reds that tint canyon and surface topography. Historical

iron ore piles and model rock stockpiles mark pit edges. Their coni-
cal and ziggurat shapes, designed to provide stability and minimize
surface footprints, give the impression of a more ancient landscape
dotted with ritual sites and primitive earthworks." Afterward, similar
comparisons to ancient earthworks would be made elsewhere in mine
country, where the chat dumps of Missouri's lead belt were described
as "the Great Pyramids of St. Francois County" and one of Scotland's
oil shale bings was rebranded as the "Niddrie Woman," a "work of land
art along the lines of ancient hill figures like the Cerne Abbas Giant or
the Uffington White Horse."[9]

It was clear that the university was interested. Like E. W. Davis and
C. O. Rost, who had lent their university pedigrees and resources to
the development of taconite processing and tailings reclamation decades earlier, the landscape architecture school would now throw itself
into envisioning a way to convert the mine-disturbed lands of the Mesabi into the legacy assets that Swearingen, Hendricks, and Rukavina
had imagined. But the U of M had one condition. The university felt
it was off mission to contract directly with private industry. However,
Hendricks remembered that they said, "If you brought a consortium of
people together, then we could be part of it, and act like an advisor." So
long as the university served the interests of the citizens of Minnesota
through an open, public entity, it could participate. In other words, a
partnership would be formed in which private and public entities col-
laborated in an open forum dedicated to the notion of reconstructing
and creating value from the postmining landscape of the Mesabi.

Back on campus, Koepke got to work. He recruited Christine Carlson to manage the project. Carlson was a research fellow and adjunct
professor with experience in regional planning for natural and cultural
resource conservation, including waterfront designs and trail planning
for the National Park Service and the Trust for Public Land. She was an
accomplished project manager and, Hendricks recalled, "understood
politics, fundraising, and community engagement"—all of which
would be important to pull off a project of this scale. Koepke and Carlson also secured a small grant that would serve as seed money for the
nascent partnership.

Throughout the summer and fall, Carlson, Koepke, and their university partners developed their ideas into a two-page preliminary

prospectus documenting a "much more comprehensive approach" for imagining the future of the Iron Range than had ever been conceived before. Its scope included not just U.S. Steel lands but the whole Mesabi Range. Its mission statement was: "In sustaining long-term resource development, [to] create well-planned, environmentally sound, made landscapes that contribute to the quality of life in Laurentian Divide Communities, both human and natural, and the economic vitality of the region and state."[10]

Carlson recalled that the big idea was "to create a larger vision for the *region* after mining" (emphasis added). In comparison to the valuable, but separate, one-off mineland reclamation and repurposing projects undertaken in the past (such as the St. James pit restoration described in the previous chapter), this approach would tackle the whole region in its multifaceted complexity. The vision would be for "a future that reclaims and restores the biological functioning and physical surface potential of the landscape, its visual quality, and its potential to improve quality of life for residents." In other words, it would embrace the multiplicity and interplay of ecological restoration, the potential to reuse minelands, aesthetics, and the improvement of life for Iron Range residents and visitors afforded by the Mesabi's mined landscape. In honor of the north-south watershed divide created by the giant granite intrusion upon which the iron formation rests, it was called the Laurentian Vision.

With the prospectus in hand, Swearingen and Hendricks got to work on building the consortium of key stakeholders to serve as the public entity the university would advise. They invited Minnesota Power, which supplied the taconite companies with the energy necessary to move the massive electric shovels, crushers, and other processing equipment. And they asked the MDNR, which was overseeing modern taconite mine reclamation under the Red Book, as well as the IRRR, which was tasked with developing economic diversity for the Iron Range, reclaiming state-owned minelands, and administering the Taconite Environmental Protection Fund. Each of these stakeholders was invited to an initial exploratory meeting with Carlson, Koepke, and the university team.[11]

So, with this who's who of mineland reclamation stakeholders assembled, U.S. Steel and the U of M presented their vision for creating new landscapes using routine mining activities. Hendricks recalled that the MDNR was "pretty skeptical to start with." And Swearingen added that the IRRR was "noncommittal." Swearingen summed up that first conversation by saying, "The enthusiasm from either party was less than what we were hoping for."

The university team was equally crestfallen. In a 2011 publication, they recalled, "Stakeholders, especially state government, utility and industry interests, were suspicious of the university and its association with any USS [U.S. Steel] agenda. . . . Several Range interests were unable or unwilling to think about the future of the region beyond mining," and "the goal setting process elucidated diverse, often opposing, points of view in perceiving and describing the Range."[12]

The initial lackluster reception and divergence of opinions are understandable. After all, the non-Native miners and residents literally named the Iron Range for its mineral wealth, and thinking about the region after mining was anathema to many who lived there. When taconite mining started, experts predicted it—and the taxes, royalties, and paychecks that flowed from it—would last half a millennium. That confidence was shaken by the 1980s downturn and the closure of Butler Taconite, yet people still primarily thought the future of the Iron Range was solely reliant on the future of active mining. But things were beginning to change. By the 1990s, the percentage of Iron Range residents employed in the mines had been declining for a decade as industries such as health care, light manufacturing, timber harvesting, and tourism gained ground. Still, the idea that mined lands could be converted into valuable non-mine-related real estate seemed foreign. Plus, state agencies are not known for their ability to turn on a dime. In fact, the methodical and consistent application of procedure is expected of a state agency and when absent (as any commissioner could surely attest) quickly comes under scrutiny by the press and agency critics.[13]

Despite the bumpy start, however, the key stakeholders agreed to meet again. They decided to invite a larger group of mining companies, other public agencies, and landowners to further explore the potential of creating postmining landforms that would leave a positive environmental and social legacy for the region.

■ ■ ■

By the second meeting, the Laurentian Vision group was starting to gain momentum. Mirja Hanson, a professional facilitator and educator from St. Paul, plied her skills to keep the group engaged and moving through some challenging conversations. Having helped mine managers and union employees hammer out contracts, Hanson had experienced difficult discussions on the Iron Range. But this was the first time that professional facilitation was employed to engage the public in a visioning exercise for the Mesabi's post-mining landscape. With her Finnish language skills she fit in with the local representatives, and her outcome-oriented and systematic presentation helped the group make decisions. Carlson and Koepke supplemented the university's in-house expertise by engaging a landscape architect and surface mine reclamation specialist—Tony Bauer, formerly of Michigan State University— who had a track record of successfully reclaiming aggregate quarries into recreational green space, housing developments, office parks, and golf courses. Bauer brought with him a slide deck to showcase projects that could serve as models for the ambitious Laurentian Vision agenda. The slideshow engaged the group's imagination, and—encouraged by Carlson, Koepke, Hanson, and Bauer—they decided to develop guidelines that would provide the foundation for a vision of Iron Range–wide projects.

One key item discussed in the early meetings was the protection of future mining resources. This issue had come to the fore when the city of Nashwauk had condemned property containing unmined iron ore resources within the city boundary to be used for a housing and commercial development, which could make future mining of this resource almost impossible. Many of the stakeholders at the meeting felt that if the iron ore was not protected from encumbrance by similar long-term plans, mining would eventually suffer from a lack of resources. Therefore, a major goal of the Laurentian Vision was to direct long-term developments away from unmined iron ore resources, thus enabling mining to continue as long as possible into the future. The group encapsulated this approach, which was not dissimilar to the mineral conservation Roosevelt and Pinchot had espoused at the turn of the twentieth century, in the phrase "sustaining long-term resource development" that opened its early mission statement.

The Laurentian Vision began to meet regularly, with Carlson, Koepke, Hanson, and Hendricks conducting check-in calls nearly every week. Thanks to Hanson, each meeting had a specific theme and agenda designed to move the project forward from concept to planning and structure to implementation. The consortium chugged through issues that were sometimes contentious because of the participants' differing priorities—such as how best to protect future minelands while accommodating the growth of communities and other industries, and how to incorporate aesthetic considerations into the reclaimed mine landscape while observing the benching requirements of the mineland reclamation rules. Around this time, the university presented the results of a graduate studio conceiving of a project named the Avenue of Mines. The concept was a linear corridor that would wind its way through the Mesabi Range and feature "Cultural Heritage Loops" to explore the historic and cultural landscapes associated with mining and miners' lives, "Connections across the Divide" to guide future mining and mineland reclamation activities to knit together the landscapes from the south to those north of the Laurentian Divide, and two "Bookends" in the communities of Grand Rapids and Babbitt. Incorporating many of the university's early landscape concepts, the Avenue of Mines was well received by Iron Range mining engineers. All of this input, developed over more than a year of steady meetings, was used to produce a road map document for the Laurentian Vision Project that was presented to and ratified by the larger group of stakeholders.

Funding for work was identified as an immediate challenge. State agencies, beholden to their annual budgets, did not initially have cash to dedicate to the process, but by this point they had expressed a willingness to spare some staff time in support of the project. The university, similarly, would contribute the time of landscape architecture professors and their graduate students, including the principal investigators, Koepke and Carlson. Carlson also obtained a second grant, this one from the university's Minnesota Design Center for the New American Landscape. The U.S. Steel Foundation and Minnesota Power provided another $250,000. With everyone contributing something—expertise, cash, or both—and voluntary participation by all, the group operated in what was later described in a publication by Carlson, Koepke, and Hanson as a "potluck style."[14]

As with other potlucks, the goodwill and trust among participants began to improve. Because good information about future mining

areas would be vital to long-range planning, the university, coop-
erating with the MDNR and mining companies, started to build an
atlas of current mines and future minable lands (defined as the subcrop of the Biwabik Iron Formation). The atlas, funded by the Legislative Committee for Minnesota Resources, was intended to serve
as a resource for land use planning that would help avoid conflicts
like what had occurred in Nashwauk. Mining companies, historically
reticent about publicly sharing their expansion plans and iron ore reserves, were forthcoming—Swearingen recalls that "all of the mines
opened up and started sharing information about where they planned
to mine in the future." While the atlas was being built, state agencies
began to see how the mineland reclamation requirements for erosion
control, drainage, and stability could be enforced while permitting
some latitude in the new designs the university was developing. All
the while, the university was bringing valuable creativity and outside
perspective to the problem. Throughout this period, Carlson, Hanson, Koepke, and Hendricks employed a leadership philosophy they
later described as "persistence, persistently applied." They repeated
and refined the Laurentian Vision's messages while working with the
stakeholders through challenge after challenge. As the stakeholders
reviewed the project concepts for the seventh and eighth times, acceptance of the ideas began to grow. It took nearly two years of regular
meetings to build the consortium's full rapport, trust, and resources.
But by the year 2000, it had turned into an ad hoc working group that
management professionals would later study as a model for a "loosely-
coupled, multi-stakeholder partnership." The consortium adopted the
name Laurentian Vision Partnership (LVP).[15]

An updated prospectus from early 2000—reflecting the larger
group's input, including that of the IRRR, the MDNR, the mining industry, and others—states that "the fundamental purpose of the Laurentian Vision is to identify alternatives for the Mesabi Iron Range as
an attractive place to live and work." Its stakeholders, representing the
major driving forces acting on the Iron Range economy, were identified as the six t's: taconite, timber, tourism, transportation, technology, and transformed communities. The group was working toward
the "thoughtful conversion of mine lands to suitable uses following
mineral depletion" that might include "public and private recreational

lakes, golf courses, parks and trails, interpretive and educational sites, private industrial parks, planned communities or hunting reserves, wildlife habitat and reforestation." This list of examples enumerated the kinds of "legacy assets" that Rukavina and Swearingen had casually discussed years earlier, only now there was a large and powerful group of mining interests, utilities, landowners, university researchers, municipalities, public schools, regional planning boards, and state agencies such as the IRRR and the MDNR behind it.[16]

During this period, the IRRR produced a foldable, one-page draft marketing piece that succinctly explained the Laurentian Vision concept as "Reclaiming the land to maximize value." It was accompanied by bullet points describing that the "Laurentian Vision is":

- A process on how to consider each potential land use in a regional, district or local environment
- It recognizes that each parcel of land on the Iron Range is unique
- A process that can paint pictures of possibilities
- A process that is open to input
- A process that encourages people, groups and communities to become involved
- A process that engages professionals in various fields to advise

As well as "what the Laurentian Vision is Not":

- A regional land plan
- A plan dictating how to deal with land planning issues
- A select group of people telling everyone else what is good for them
- A "secret plan" devised by mining companies, DNR and IRRRB . . .
- A preconceived plan that locks up land, keeping it from being used in the future.

To stir the imagination, the document had a hypothetical before and after image of a water-filled mine pit that had been reclaimed into a marina accompanied by three-story waterfront condominiums atop a gently sloped and revegetated shoreline.[17]

Just a few months later, the Iron Range was rocked by the announcement that the Erie Mine and Processing Plant in Hoyt Lakes (where Dickinson and Youngman had conducted their pioneering reclamation experiments) would be permanently closed. The Mesabi's second taconite plant closure injected fear for the future and a new sense of urgency about what mining communities might face after the shovels stopped digging. Against this backdrop, the LVP was proposing an idea that sought to "create resilience against the boom or bust economic cycles" and "to use the mining process to reconstruct a landscape capable of generating its own sustainable economy."[18]

The group suggested that perhaps the Mesabi giant could give life to the region's economies in ways beyond the mineral value alone—through the lakes and landscapes created by mining activities. Koepke,

Mine closure rocks Range

Mine closing will idle 1,400 workers; devastate community

HOYT LAKES (AP) — Now that LTV Steel Mining Co. has decided to close its taconite plant here, eliminating 1,400 jobs, residents worry that the town's future is bleak.

"In a nutshell, this is the most devastating thing to ever happen to this town," said Doris Kopp, who works at the local grocery.

"It's going to affect everybody," Kopp said. "It's scary. It's very scary. We've had hard times before and survived, but this time, I just don't know."

The mine, which once employed more than 3,000 people, is the town of about 2,400's largest employer; the next largest business employs 42 people.

The plant is being closed because ore quality has declined, making it too expensive to refine the low-quality taconite pellets at the outdated Hoyt Lakes plant, said LTV Steel president Richard Hipple on Wednesday.

LTV Steel Mining will suspend stripping operations — removing layers of soil and rock that cover the taconite ore — effective Sunday, and the first 120 workers will be laid off by the end of August, Hipple said.

Mining, crushing, pelletizing, and shipping operations should continue for another year. Hipple said 1,280 workers would be kept on the job through the middle of 2001.

"It's important to note that this is not a people issue. Nobody is at fault," Hipple said. "Everyone has tried hard to do the best job possible. In fact, these efforts have extended the life of this mine."

The decision will have a major

'It's important to note that this is not a people issue. Nobody is at fault. Everyone has tried hard to do the best job possible. In fact, these efforts have extended the life of this mine.'

RICHARD HIPPLE
LTV Steel President

impact on the Hoyt Lakes area, said John Swift, commissioner of the Iron Range Resources and Rehabilitation Board.

"The jobs can pay between $65,000 and $70,000 a year, and now those jobs aren't going to be there," Swift said.

Gov. Jesse Ventura said the state will do whatever possible to help with the transition.

"Nobody likes to see companies close down and leave but that's the negative part of doing business," Ventura said.

Ventura, Swift and Jerry Carlson, commissioner of Trade and Economic Development, met with LTV officials Wednesday in Duluth to discuss the closing.

Sen. Paul Wellstone said he was "deeply saddened" by the announcement "and angered that the closure is due at least in part to cheap steel and iron ore being dumped on the U.S. market from places like Russia and Brazil. We need tougher enforcement

See • Closing / Page A16

MARK SAUER

An ore train hauling taconite pellets leaves the LTV Steel Mining Company plant in Hoyt Lakes Wednesday. The plant will close for good by the middle of 2001.

The announced closure of Erie Mine and Processing Plant in Hoyt Lakes was a shocking turn for Iron Range residents. *Hibbing Daily Tribune/Mesabi Daily Tribune*

Carlson, and Hanson later reflected that, "A new era of lakes and land-scapes was emerging in the region" that had "opportunities for recon-necting natural systems, strengthening community relationships, and meeting the university's larger educational goals." They recognized that earlier mineland reclamation practices, or lack thereof, "chal-lenged the creation of a positive landscape legacy for future genera-tions in mine country." But their aim was to "help the region retain its role as a productive and healthy landscape that provides a high quality of life in northeastern Minnesota."[19]

To paraphrase, the LVP wanted to capitalize on what they called "free energy materials" that were created by active mining practices (such as overburden, lean ore, and waste rock) and naturally occur-ring reclamation (such as Gilbert Leisman's volunteer birches, aspen, and poplar) to consciously shape the pit lakes and mined landscapes. The lakes and landscapes would reconnect natural pathways severed by historic mining practices (such as creating vegetated wildlife corri-dors across the iron formation) and provide a platform for future de-velopment of the lands after mining was complete. The Iron Range communities could participate in planning for these transformations over time, and the mining companies could build in-house planning and reclamation design skill sets to maintain their personnel, reduce consultant costs, and continue land shaping during peaks and valleys in the industry. Transforming the mined landscape would improve the quality of life for residents and serve as an economic engine for mining towns that could help counterbalance the cyclical economy of resource extraction. A systematic approach to transformative projects, with full engagement from the "six t's" and regulatory agencies, would mean fewer hurdles along the way (for example, regarding compliance with rules, acquiring funding, and so on). And by openly discussing these challenges and opportunities in a public forum, the LVP could both inform ad hoc members and engage them in developing solu-tions, while broadcasting to other Minnesotans the importance of the work. The project would introduce a new vocabulary to describe the region's future and to facilitate a better understanding of and deeper enthusiasm for long-term change.

■ ■ ■

It was an inspiring mission almost as big as the Iron Range itself, and the LVP was ready to put it to the test by beginning some high-profile projects. The theme for the eleventh meeting of the LVP in December 2000 was "Launching Project Deliverables." An estimated $650,000 in cash and in-kind donations had been spent getting to this point, and the project wanted to "get the 'boat pushed away from the dock' and *show* the region some concrete images about regional development."[20]

The LVP's first highly visible action—a public design workshop— was set to take place in Virginia, Minnesota, in 2001. This area at the heart of the Iron Range encompassed the greatest number of active taconite mining operations: U.S. Steel's Minntac Mine, Ispat Inland's Minorca Mine, and Eveleth Taconite's Thunderbird Mine. The process started in the spring with a public meeting sponsored by the Virginia Community Foundation. LVP leaders met with the mayor of Virginia, mining and city officials, and the public to discuss whether the city would host an intensive multiday workshop focused on conceptual planning and design of alternative solutions for transforming mine-lands into usable, healthy landscapes—and, if so, to identify the ideal site for this first exercise. Carlson remembered, "They were game; they had sites around the city of Virginia that were not doing anything and had to be redeveloped." Jointly combing through the possibilities, the stakeholders selected a site of some five hundred acres located east and south of town, consisting of a complex of conjoined, idled natural ore pits, including the Missabe Mountain, the Commodore, the Shaw-Moose, and the Rouchleau. This site not only was well situated for redevelopment but also served as the location of the city of Virginia's water supply and Eveleth Taconite's future mining area. Thus, it seemed like the ideal spot that brought together all the elements of the LVP's mission—community development, environmental protection, economic diversification, and protection of lands for future mining.

Koepke and Carlson called the intense, multiday public design workshop a charrette—one of the first words in the new vocabulary the LVP introduced to the Iron Range. This specific use of the term, which means "cart" in French, dates back to nineteenth-century Paris, where architecture professors would send a cart to pick up students' design work at the submission deadline. It is still used to define "the intense final effort made by architectural students to complete their solutions to a given architectural problem in an allotted time or the period in which such an effort is made." But it also commonly means "a meeting [really, a structured, intense, multiday workshop] in which all stakeholders in a project attempt to resolve conflicts and map solutions." As it turned out, both definitions accurately depicted the whirlwind of events and spontaneous design work that took place in Virginia over the intense half week of October 10–13, 2001.[21]

It was a Wednesday afternoon when Koepke, Carlson, and the other landscape architects arrived. Some, such as Tony Bauer from Michigan, flew in, but most of the university contingent drove up. Altogether, there were a dozen landscape architects, ranging in experience from national experts to graduate students. As soon as they arrived, Koepke, Carlson, Bauer, and a mining engineer from Eveleth Taconite introduced the LVP's mission, background, partners, and expectations for the charrette. Regional experts gave the group an immersive course in the natural history, cultural history, and current affairs of the Iron Range, and then Carlson presented the issues and opportunities for the charrette. It is noteworthy that some of the experts presenting that day were from state agencies whose workers were officially striking at the time of the charrette; nonetheless, they volunteered their time to make this important project happen, demonstrating the level of personal commitment state officials had made to the partnership by then. In the late afternoon, participants were given a tour of the mines and the communities within the design limits set for the charrette, and over dinner, they were briefed on the mining process.

The landscape architects were, as Swearingen recalled, "from all over the country," and none except for Bauer and James Pettinari, who hailed from the Iron Range city of Buhl, had ever seen the Iron Range or an active mine. By 8:00 pm the first day, the charrette officially kicked off with a public open house. Bauer gave a presentation on "Shaping Landscapes," and LVP leaders summarized the overall intent and goals of the next few days of work. During the open house, everybody—from mine representatives to city officials to regional businesses to the local populace—was invited to ask questions and share their visions of what would be best for the area's future and the landscape of the Mesabi. The landscape architects took notes and gathered ideas. They had been divided into three teams of four people, with each team dedicated to a certain aspect of the future visioning process: community development and housing, recreation and open space, or economic development. Carlson and Bauer were the overall process leaders.

After the initial public meeting wound down, the three teams, in Swearingen's recollection, "literally worked around the clock" to create sketches of how all the ideas could meld into the landscape. Pettinari, in particular, was noted for his "expert drawing skills that could quickly convert the discussion ideas into exciting concepts on paper." Eveleth Taconite provided engineers for up to twelve hours per day to advise the teams—telling them what was feasible and what was not in terms of general engineering and mine development. Other resource experts were brought in at intervals, including the mayors of Virginia, Eveleth, Mountain Iron, and Gilbert. Unlike an engineering project that is undertaken with a single-minded focus to solve a particular problem, the charrette tried to envision multifaceted solutions by mixing together the creative output of all three of the design teams into a single plan.[22]

On Friday afternoon, a second site tour was conducted to begin validating the landscape architects' concept plans. By that evening, contributors were reassembled in a second public open house to view the progress the teams had made and to provide any final input or refinements. Hendricks recalled, "People would come in and watch. [The teams] were drawing; they were kicking ideas around. They were putting up sketches. It was fascinating to watch the whole thing."

By noon on Saturday, the group presented the final products to all of the project's stakeholders, who included, in Carlson's words, "key community leaders, citizens, mine engineers, movers and shakers, and national design and planning experts." The whole process concluded with a "farewell social" and a press conference. Everyone was included. Hendricks recalled that the stakeholders' reactions to the drawings presented at the final open house were overwhelmingly positive: "People who watched the presentations were like, 'Wow, we never thought you could create this.' An access ramp into the pit where you could have marinas, boat launches, and walking trails? Or a business district down there that did not impede mining?"

Iron Range native James Pettinari sketched enticingly scenic possibilities during the Virginia charrette. *Courtesy of Northwest Architectural Archives, University of Minnesota Libraries*

Importantly, the professional facilitation techniques employed during this and future charrettes allowed those involved to air out some vexing issues regarding the future of the range cities that had not previously been brought to light. Carlson said the events provided a neutral territory to present and discuss conceptual solutions to these issues. For example, the Virginia charrette made it clear that US Highway 53, linking Eveleth to Virginia, would need to be moved to accommodate expansion of the Thunderbird Mine. Swearingen recalled, "It was the Quad Cities charrette that started this discussion and laid the groundwork for how to relocate Highway 53." Rather than shy away from the potential highway relocation, the charrette made it a centerpiece for discussion. One landscape architect team was dedicated to developing concepts for the "Gateway Bridge" that featured links to the paved Mesabi bike trail and an overlook tower. Sketches showed the bridge bypassing an island dedicated to wildlife habitat and connecting into

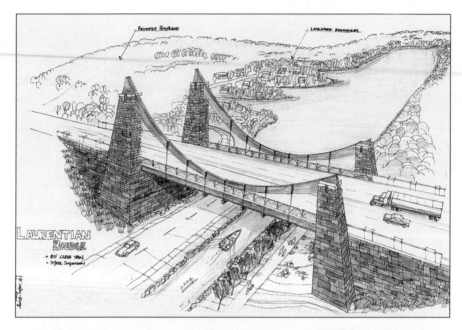

The Virginia charrette prominently featured multiple "Gateway Bridge" concepts designed to allow for mine expansion, inter-community connection, and enhanced recreational and cultural opportunities such as incorporating the Mesabi bike trail and an overlook tower. *Courtesy of John Koepke*

a lakeshore housing development and public park. The landscape architects were painting a picture of the future that featured the highway relocation as a potential hub of activity, like the Golden Gate Bridge, rather than a potentially contentious mine siting issue.

Other parts of the Virginia charrette drawings featured urban ecology corridors that offered greenways for wildlife to cross the iron formation from north to south, subaqueous habitat zones created by programmatic plantings as the water levels rebounded after mining was completed, and a viewing corridor into the future pit lake known as the "Iron Gap." There were natural amphitheaters, ski and bike trails, and what was termed a progress park for new businesses.

According to Koepke, Carlson, and Hanson, the charrette drawings were based on a "working hypothesis that mining companies can shape landscapes with certain mine activities in strategic locations across the Mesabi Range to improve the region's cultural, economic and environmental life." The landscape architects were trying to show "a vision of the region as a nonextractive landscape"—in other words, what the

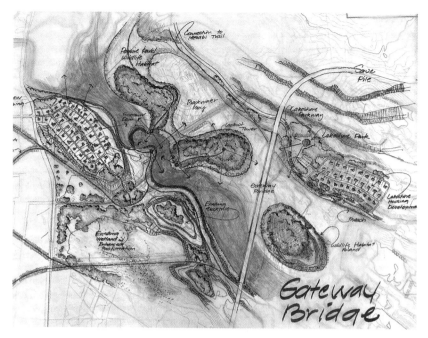

The Virginia charrette included discussion of a "greenway," or wildlife corridors across the iron formation. *Courtesy of Northwest Architectural Archives, University of Minnesota Libraries*

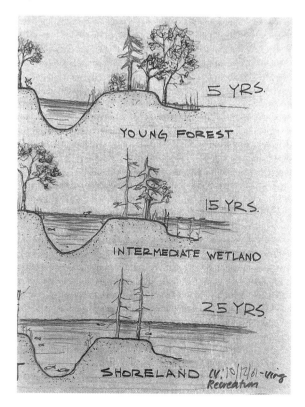

A charrette drawing showed how purposeful plantings could be used to develop shoreline habitats as water levels in mine pits changed over time. *Courtesy of Northwest Architectural Archives, University of Minnesota Libraries*

Iron Range could be when mining was over. And the charrette drawings were used to depict the idea that the machinery of today's mining could be used to shape the minescape into "attractive, productive land, using thoughtful site planning and visual design devices." The whole idea was eventually boiled down into what would become the LVP's catchphrase: "Pits and Piles into Lakes and Landscapes."[23]

The first charrette was a huge win for the Laurentian Vision Partnership. It sparked development of a Quad Cities Land Design Planning Partnership, which comprised thirty-nine members representing a broad spectrum of public, private, and nonprofit agencies that eventually worked through a timeline of actions for relocating Highway 53 without impeding the progression of mining. As they engaged in this follow-up partnership, the members reflected on the charrette process. A representative of the city of Virginia said, "We have started something good and must continue it. The City of Virginia is *in*." A representative of Eveleth Taconite said, "The mine is always creating new land forms. As long as costs are not increased, we are interested in knowing the best ways to stockpile material for future users." And a representative from the city of Eveleth said, "I compliment the collaboration of the group up to now. This is an exciting opportunity to plan for the future." Through these testimonials, it seemed as though the messages LVP was trying to convey—community engagement, transformation of the future landscape through mining, collaboration, and long-range planning—were being received.

■ ■ ■

This public success helped the LVP to gel and to attract new members. Other Iron Range communities approached Carlson with ideas for other charrettes, such as a Laurentian Lakes Project proposed by the city of Buhl and a Canisteo Design Charrette for Greenway's mine sites on the western Mesabi, some of which presented geotechnical and flooding issues for the cities of Bovey and Coleraine. Koepke and Carlson later reflected that during this period, the LVP became established as "an institutional framework" and a "sustainable decision-making resource on the range." Partnership meetings were scheduled for three times per year. By the end of 2002, the amount of cash and in-kind contributions to the project had risen to nearly $780,000. The LVP used this momentum to begin further outreach and identify the sites of additional design charrettes.[24]

The second charrette was set for May 2003, this one focusing on the Chisholm-Hibbing area and encompassing the minelands of Hibbing Taconite Company as well as the numerous idle natural iron pits, stockpiles, and tailings basins of the Central Range. Whereas the Virginia charrette had focused on infrastructure needs, the Hibbing-Chisholm workshop delved into possible solutions for land use conflicts, such as moving a highway, creating a corridor for nonmining commercial and industrial uses, and redeveloping waterfront minelands for cabins, parks, and wildlife habitats. The team followed the same type of high-energy, engaging, and visually stimulating design process, with local mining engineers providing guidance. Some participants who had been intrigued by the first event in Virginia traveled halfway across the Iron Range to participate in the second. Over ninety businesspeople, public officials, mining company representatives, and citizens (not including the university's teams) signed on. Video recording by a film crew documented the charrette design process. (The result, on VHS, is still stored at the Minnesota Discovery Center.) And follow-up work on the design ideas discussed during the charrette process was spearheaded by a nonprofit group of local interests known as the Central Iron Range Initiative.

In 2007, the LVP held a third charrette for the Biwabik and East Range area. In contrast to the previous two, which had focused on lands that had been disturbed by mining activities in the past, this one centered on the landscape design opportunities presented by the expansion of an existing taconite mine into previously undisturbed lands. It included the preservation of wildlife migration corridors amidst the future mining operations. And it proposed concepts for community expansion along a highway redevelopment through town as well as recreational trail connections from the east range cities up to Giants Ridge. In an educational video, an ArcelorMittal mining engineer described the landscape design elements that were incorporated into the new mine's layout, using some of the vocabulary that the LVP had introduced to the Iron Range.

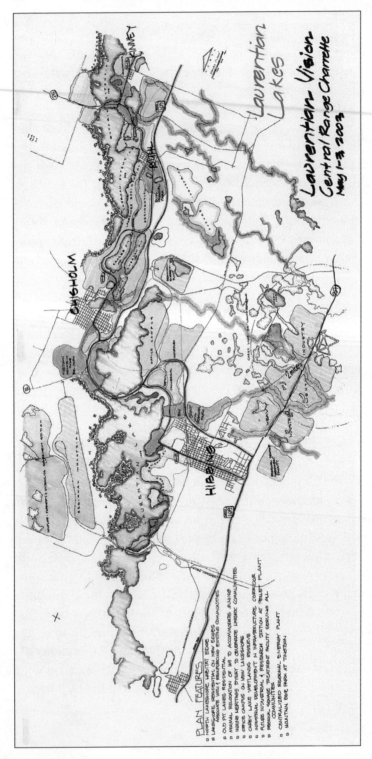

The Hibbing-Chisholm workshop considered solutions for land use conflicts, such as rerouting a highway. *Hibbing Chamber of Commerce*

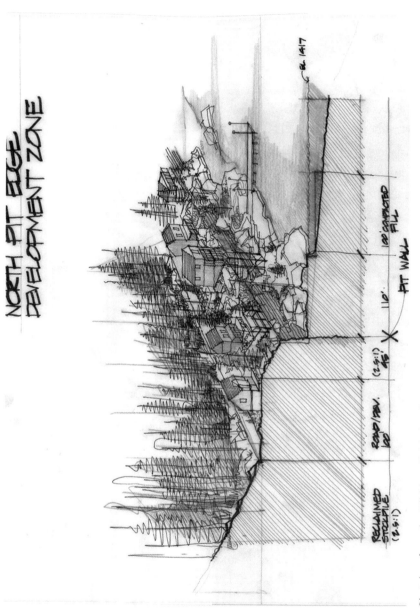

Sketches from the Hibbing-Chisholm charrette suggested developing waterfront residential properties along the edge of a mine pit reclaimed according to the mineland reclamation rules. *Drawing by John Koepke*

FUTURE LANDSCAPE

Giants Ridge

Biwabik

McKinley

A third charrette drawing envisioning Laurentian Vision concepts incorporated into a new mine in the Biwabik area. *Courtesy of John Koepke*

All three charrettes engaged the public in the mine development and reclamation process in a way that had never been done on the Iron Range, resulting in revelations for both the mining companies and local residents. Koepke, Carlson, and Hanson wrote that the events "changed the thinking of mining companies about a mine's end use and the companies' role as a community partner. It also changed community attitudes toward mining companies." They praised the work of the local mine engineers: "Participating mine engineers demonstrated strong problem-solving skills, good design sense, and the capacity to be team members capable of addressing community issues. Hence, their stature within participating communities increased."[25]

Swearingen and Hendricks also emphasized the value of the university's visionary contributions. In Swearingen's opinion, the university brought the creativity, design skills, and outside perspective that was needed to capture the charrette participants' imaginations. He also noted that the "U of M was doing it for nothing. They were doing it for their costs." He recalled that each charrette cost only $40,000 to $50,000 to produce, which to him felt like a bargain for the tremendous professional value that was brought to each session. For what seemed like peanuts, Swearingen's, Hendricks's, and Rukavina's visions now had a set of concept-level blueprints.

■ ■ ■

In the mid- to late 2000s, with the success of the charrette program established, the LVP set itself on a mission of educating the public and becoming an institution, not just an ad hoc working group. Chris Carlson prepared a charrette workbook to instruct future leaders on how to work through the process, including details such as mock agendas, room layouts, and roles for each of the players. By this time, the IRRR recognized the multifaceted value that LVP could bring to the region in support of the mining industry, community development, and sustainable reclamation of mined lands. The IRRR created a logo, developed a website, and hired Jim Plummer, whose background in urban and regional land use planning would be employed as the paid, part-time LVP coordinator. The IRRR also established an innovation grant program to provide matching funds for mineland reclamation projects that met the LVP mission. Regional subgroups in the East Range, Quad Cities, Central Range, and western Mesabi worked on the community level and reported their progress to the LVP. The MDNR provided maps and tools for the partnership and the public related to land use planning, underground mine mapping, and hydrology studies. And Koepke, Carlson, and Bauer conducted a three-day workshop at Hibbing Community College to educate mining engineers on reshaping the landscape at various scales and times during the mining process. "Land Design Opportunities in Taconite Mining: A Land Shaping Workshop" included the rationale, landscape design principles, templates, and team exercises for conceptualizing LVP-inspired ideas into active mine planning as well as how to engage the public and state agencies along the way. The goal was for the LVP to instill in-house design and reclamation skills at the mining companies. Forty senior and junior mining engineers attended the course, as did MDNR mine reclamation specialists and IRRR staff. These efforts raised awareness about the LVP and encouraged the participants to think about "how mine features [pit walls and edges, stockpiles, haul roads, terraces, and so forth] could be shaped into attractive, productive land . . . using thoughtful site planning and visual design devices." Koepke and Carlson were, of course, natural educators who stoked attendees' enthusiasm. Koepke remembered, "The course ratings were all A+."[26]

In the enthusiastic atmosphere of the land-design workshop, mine engineers applied their newly acquired skills to some highly visible stockpile areas where the LVP concepts could be implemented in the real world. These projects would reshape and vegetate existing stockpiles to enhance their habitability for wildlife and visual appeal for residents, an undertaking that had rarely entered into earlier mining practices. In Virginia, one of the most public-facing stockpiles (United Taconite's Stockpile 1406) was rebuilt to "undulate the ridgeline to create topographic variation, soften or break manufactured terraces, decrease planarity of slopes, add texture and depth to surfaces and vary planting patterns." These complex design ideas were given bold and simple concept names, such as "the birch wedge," "lobes and valleys," "the bands," "the seasons," and "pines against the horizon." Rough grading to reconfigure the slopes was done in September 2009, and cover crops, wildflowers, shrubs, and trees were planted. Koepke remembers that good fortune smiled upon the plantings by dousing them with enough rain that the next summer's inspection found the ground blanketed with black-eyed Susans and a diverse host of trees and shrubs that had lived through their first winter. An MDNR inspection in 2012 found the plantings to be in good health, with knee-high tamarack, birch, and dogwood as well as house-high aspen, proving that a $98,870 investment employing LVP's techniques could grow purpose-planted trees while meeting the vegetation standards embodied in the Mineland Reclamation Rules. It was the first project funded by IRRR's innovation grant program.[27]

Next, the Hibbing Taconite Company and the university team designed "Grouse Mountain," in which careful grading techniques implemented by mine operators were complemented with grouse-favored plantings of dogwood, highbush cranberry, nannyberry, Juneberry, crab apple, and white spruce. The mine operators, who were mostly sportsmen, loved the idea and challenge of creating purpose-built grouse habitat, and grade school children from Hibbing were engaged to help with nearby plantings. Shortly afterward, in a third innovation grant project, Keewatin Taconite's Dump 43 was designed to maximize both its visual appeal and site stability using plants that varied seasonally in color and bloom while being able to self-seed and adapt to drought and climate change conditions. Each of these projects provided a visible demonstration that LVP concepts could be realized for prices ranging from $82,000 to $683,000 to provide aesthetically appealing and ecologically productive landscapes populated by carefully selected and planted native species. They also demonstrated that IRRR's new innovation grants could provide tangible short-term results.

By 2011, the LVP seemed established. Charrettes had been held for most of the active taconite mining areas of the Mesabi. Mining engineers and civil servants had been educated in the partnership's vision and tools. And for thirteen years, it had maintained its "loosely coupled" partnership and regular meeting schedule. Meetings often had forty, fifty, or sixty participants and ran the gamut of topics related to the future of the Iron Range. IRRR had hired Plummer to professionally coordinate LVP activities, and all kinds of people were volunteering their time to contribute to LVP: the MDNR, the mining companies, cities, regional joint powers boards, engineering and environmental consultants, and general citizens. Later that year, the university landscape architects published this assessment: "The partnership is now a legitimate forum for discussion of a variety of issues related to mining and community development."[28]

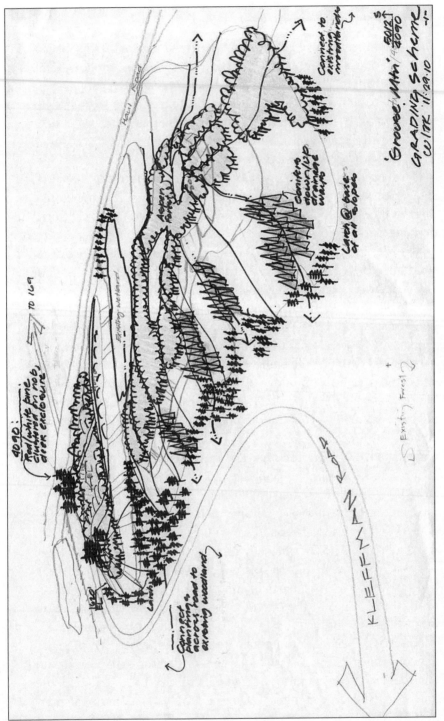

A sketch of the proposed Grouse Mountain showed how the landscape could be reclaimed for game bird habitat. *Courtesy of John Koepke*

Students from Hibbing helped plant trees that would favor grouse habitat, 2015.
Author's collection

Having established LVP as an Iron Range institution and legitimate forum, in 2011 Plummer and a dedicated group of participants spearheaded an effort to create a formalized strategic plan—a blueprint for the ways in which LVP would dedicate its time and resources for the next decade. A major strategy of the 2011 LVP Strategic Action Plan was to expand the use of IRRR's innovation grant program to construct more projects, like the stockpile demonstration projects, that produced visible results over short- to mid-term time frames. These were commonly referred to as "bricks and mortar" projects to emphasize the tangible outcomes that were being sought over the decade to come.

■ ■ ■

In 2013 and 2014, the LVP embarked on what Koepke described as its "penultimate project" and Carlson called the "apex of LVP's work." They, Hanson, Julie Jordan, and other MDNR officials and landscape architects and designers engaged with Northshore Mining Company to develop an installation field manual for the Peter Mitchell Mine Landscape Framework Plan. The purpose was to promote and execute a full-scale master plan for an operating taconite mine that incorporated all of the LVP's principles and techniques. The document laid out a mine-wide reclamation framework that would systematically transition Northshore's active mine pit into a diverse lacustrine (lake) habitat modeled after the shorelines of Lakes Superior and Vermilion. The transition would take place as the ore reserves were removed and the Peter Mitchell Mine developed from east to west over a period of about sixty-five years. It included a "living laboratory" where innovative reclamation concepts could be tested and perfected and a nursery to harbor soil and vegetation that either grew naturally in the area or needed to be cleared for mine progression—capitalizing on the opportunity to use "free energy materials," as Koepke championed. Also included were a visitor interpretation area; development of upland forest, grassland, and sharp-tailed grouse habitats; littoral zones and deepwater fish habitats along a human-made environment called the "Finn Fjord"; and nonmotorized public access to water bodies and trails. A small companion field guide for mine operators was printed and bound small enough to be tucked into the visor of a shovel, bulldozer, or haul truck. And the appendices to the plan included recommended native seed mixes, design worksheets, and other materials to enable hands-on decision-making. All of these concepts were developed in the course of four intensive eight- to twelve-hour work sessions. The project was funded by a $432,600 innovation grant from IRRR that was matched by private dollars. Ultimately, the MDNR and Northshore Mining Company agreed to include the plan in the facility's permit to mine. At one point, it was presented to Mesabi Trust, the group of mineral rights owners who leased the lands of the Peter Mitchell to Northshore Mining, one of whom Koepke recalls having brought up the idea of creating an industrial-recreational-ecological state park after the mine was closed. In May 2015 the American Society of Landscape Architects awarded the work documented in the plan both a Communication Honor Award and an Analysis and Planning Merit Award.

The Northshore Mining Company framework plan envisions a future lake habitat inspired by the shorelines of Lake Superior and Lake Vermilion. *Courtesy of John Koepke*

But even a celebrated and visionary public-private partnership cannot stem the tide of time. Retirements and attrition began affecting the LVP; Swearingen retired after the second charrette was completed, having served thirty-seven years in the steel industry, and Hendricks retired in 2016, having worked for forty-two years in the mining industry. Funding for the university's involvement ran out, leading to the loss of the landscape architecture faculty and students. Soon, none of the original visionaries from the 1998 kickoff meeting remained. When the university left, it was unclear whether intellectual property issues would prevent the continued use of the charrette materials, education plans, and other work products. Therefore, when the mining engineers who had been trained by the LVP moved on with their careers, the resources were not available to train the next generation in the LVP's original ideals or methods. As of the time of this writing, all of the original charrette leaders, landscape architects, and engineers have either retired, disengaged from LVP, or changed jobs.[29]

On May 2, 2017, however, Koepke, Carlson, and the other University of Minnesota landscape architects donated the original LVP project materials to the Northwest Architectural Archives at the University of

Minnesota Libraries. Koepke and Carlson worked with the archivists to develop and design an accurate project narrative, chronology, and composition of drawings, strategy, and documentation of community participants. The dedication began with a three-month display of the LVP's documents at the Elmer L. Andersen Library on the University of Minnesota campus. At the opening of the exhibit, a crowd of hundreds, including university regents, attended to mark the project's success, to see the charrette drawings, and to hear the original project leaders reflect on its goals, visions, and achievements. It was certainly a celebration, with wine and hearty appetizers and the mild sunny weather of the end of the school year, but to some it also represented the beginning of the end—or at least the end of the beginning—of the LVP.

■ ■ ■

Three years later, in 2020, after nearly a decade of operations under the 2011 strategic plan, the IRRR conducted a second strategic planning process to help the LVP chart its direction for the next five-year period. One outcome was a change in the name, from the Laurentian Vision Partnership to the Mineland Vision Partnership, or MVP. A new logo was created, and the group's mission and vision changed. The MVP's current mission statement is "Shaping evolving landscapes for future generations," and its goal is now to "develop opportunities for dynamic minescapes, preserve lands to sustain current and future mining and provide resources and education." This new name, vision, and mission—reminiscent of Roosevelt's plea to southern audiences during his 1905 rail tour with Greenway, when he asked listeners to work not only for their own well-being "but for the well-being of the generations yet unborn"—reflects a change to a new generation of leaders and participants in the organization.[30]

During our 2021 interview, Swearingen expressed some misgivings about the trajectory and long road ahead for MVP. In his opinion, it had "drifted off into all kinds of things" and gotten "way too broad based." He reiterated that the original intent was simply to address the question "How do we develop iron mining in northeast Minnesota to leave something of value—no, something *more valuable*—behind?" And, with the charrettes for creating legacy assets in hand, to direct funding to make sure that such blueprints were being followed. But in his opinion the idea had become complicated, and people were losing focus.

Maintaining focus over the timeline of the LVP/MVP is indeed a formidable challenge. In the mid-1990s, around the time LVP started, some people estimated that because iron mining had been conducted on the Mesabi for more than a hundred years, it would take a second hundred years to reclaim the lands that had already been mined. Add to that another century of additional reserves for many of the taconite operations, and we have a time frame that far exceeds human career spans and motivation. How can we sustain the vitality of a loosely coupled, multistakeholder, all-volunteer project for a hundred to two hundred years? Plus, there is constant uncertainty about the global market for steel. While globalization and reduced production has decreased the world's reliance on the Mesabi's iron, the supply of and demand for iron and steel varies daily and can hinge on unpredictable events (such as the closing of Brazil's iron mines following a series of tailings basin failures that occurred in the late 2010s). These swings can influence which rock is considered merchantable ore and which is simply waste rock. Mines almost never close because they have completely run out of ore; rather, they close because the remaining ore is not cost effective to recover, a condition that may shift when the global economy changes. The result is a lack of clarity and agreement about when any site is truly "postmining" and ready for repurposing to begin. So, facing an uncertain starting line and a goalpost that may be sixty-five to two hundred years in the future, plus or minus a significant margin of error, isn't it natural that the attention of the human beings involved in the MVP would begin to wander to more immediate issues? It is a problem like climate change or retirement savings—one requiring constant contributions to make meaningful progress but having such a long time frame and enough uncertainty that it is difficult for humans to sustain the necessary enthusiasm to stay engaged.[31]

Koepke and Carlson foresaw this challenge when they identified that "the recognition and management of the Laurentian Vision's long timeline in the context of more immediate needs and aspirations of multiple partner agendas" would be difficult. Swearingen speculated, "It might be right that the long time horizon has played into the loss of interest." He went on to say that, although the LVP existed in name and mission for more than twenty years, in comparison to the project's centuries-long time frame, "it was still in its infant stages. It was just really starting to cook." Despite his concerns, however, Swearingen finished our interview hopefully, stating, "Some people are probably still paying attention to what came out of the charrettes."[32]

Indeed, some are.

. . .

It was a picture-perfect October day: one of those rare occasions on the Iron Range when the sun warms the stones, the mosquitos have retired for the year, and the breezes are gentle enough that the leaf stems remain attached to the autumnal tree branches. I was imploring the driver of a luxury coach bus to trust me—the view would be worth it— as we made the treacherous descent down a gravel switchback to the maintenance causeway beneath Minnesota's tallest bridge. In the bus seats behind us, the excitement was mounting as forty up-and-coming young business leaders, brought to the Iron Range as part of the Minnesota Chamber of Commerce's Leadership Minnesota series, prepared their cameras. The young businesspeople had just toured U.S. Steel's Minntac facility and were now listening to me speak through the bus microphone about the roles of engineering and mineland reclamation in the economy of the Mesabi. As we neared the water level, I quit speaking to let the landscape take center stage. The heft and grandeur of the brand-new piers were impressive as they soared overhead to support the deck of the bridge 204 feet above. The backdrop was vivid stripes of blue and red, like a flag, sprayed with the yellows, oranges, and browns of the tree leaves on the pit edge above the water and below the clear sky.

We were at the bottom of the Shaw-Moose open mine pit. Overhead was the brand-new Thomas Rukavina Memorial Bridge that connected the city of Virginia via US Highway 53 to Eveleth, Gilbert, and points beyond. It was 2018, one year after the original LVP materials had been dedicated to the Northwest Architectural Archives, as we stood beneath the embodiment of what the LVP landscape architects had envisioned as the Gateway Bridge seventeen years before. It had dedicated lanes for bicycles, pedestrians, and all-terrain vehicles. It was viewable from Virginia's new—and aptly named—Bridge View Park. And while most of that day's visitors were more entranced by the view than the history, the bridge stood as a manifestation of the LVP's ideals of preserving minelands (allowing the Thunderbird Mine to continue its northward progression), developing community (Vir-

The Thomas Rukavina Memorial Bridge, the first independently funded, full-scale construction of a Laurentian Vision concept, connects the city of Virginia via US Highway 53 to Eveleth, Gilbert, and points beyond. *Courtesy of Ardy Nurmi-Wilberg*

ginia now hosts Bridge Daze at the new park every year), and economic development (representing a $240 million state investment in the future of the Iron Range). Aside from the initial demonstration projects, it was the first independently funded LVP charrette concept to be constructed.[33]

The day ended with a group photo as the young chamber leaders clustered at the bottom of a 120-year-old natural ore mine pit beneath the gleaming new bridge. Only the bus driver vowed never to come back.

■ ■ ■

In addition to the bridge, many smaller LVP projects have come to fruition. The IRRR has funded forty-one other enhanced mine reclamation projects to date, a total of $5,116,000, through its Minescapes Grant Program (formerly the innovation grant program), including Black Beach in Silver Bay and the world-class disc golf course in Buhl. MVP continues to hold three annual, professionally facilitated meetings where topics related to the use and development of minelands are presented at workshop-style gatherings. With an average attendance of fifty, interest and participation remain high for an organization over two decades old. On the Iron Range, where both communities and the mines that surround them are seeking to grow, the MVP remains a forum for discussion, education, and access to planning resources such as maps and other educational tools. Through these actions, the group seeks to moderate land use discussions and, ideally, prevent planning mistakes with potential billion-dollar consequences while promoting well-considered mine reclamation and repurposing projects.[34]

Though they have personally moved on from the organization, Swearingen and Hendricks believe the mission and actions of the LVP retain the potential to inspire a new generation. Indeed, Swearingen emphasized, "Dennis and I—our hearts are in it—we both saw the value in it—and I will tell you, it was exciting stuff." They believe the LVP story, which embodies both a sound business strategy and a dedication to environmental and social responsibility from the mining industry, has the potential to change people's opinions about mining. In Hendricks's words, "Especially now—people look at mining and ask, 'Do we really need it? It is just a dirty business.'" But those people could look at the mission of the LVP and recognize that "everything coming out of this was nothing but good."

While the initial aims were simply to solve an economic problem for Minntac, what Swearingen and Hendricks grew to value most from LVP was the public's engagement in the future of the Iron Range. Hendricks closed the interview with a reflection on the LVP legacy: "It brought exposure to the local populace about mining—where it is going—and engaged them in a way that they had never been engaged before. How do you measure that? I think it is invaluable." Swearingen and Hendricks, having retired from the mining industry, are now part of Hendricks's "local populace" who are choosing to live out their lives on the Iron Range. The same is true of Dave Youngman and Dan and Julie Jordan. Sam Dickinson lived on the Iron Range his whole life. Based on this small sample, it seems that modern-day mine reclamationists tend to stay on the Mesabi.[35]

While Hendricks values the groundbreaking public engagement, Dan Jordan and Julie Jordan—who participated in critical elements of the LVP during their careers at the IRRR and the MDNR, respectively—emphasize the word "partnership." Never before has a genuine public-private partnership succeeded in drawing in so much interest and voluntary participation. All the mining companies are involved, as well as state agencies, cities, counties, consultants, and the public. People dedicate themselves to the open process, at times with enough professional risk that, Dan Jordan recalls, "Those charrettes were built on controversy." The strong partnership ethos is still reflected in the broad, voluntary participation that is a feature of every MVP meeting.

Carlson and Koepke, perhaps reflecting their professorial backgrounds, emphasize the importance of the education the LVP provided—not just to mining engineers and MDNR and IRRR officials who attended the design workshop but to the general public—about the value of the revitalized landscape as the true bedrock and infrastructure of the Mesabi Range upon which any future for Iron Rangers will rest. This education, once received, never really goes away in the minds of the informed, and in this way the LVP's impact lives on in the receptiveness of miners and officials and residents to new ideas.

My personal favorite legacy of the first act of the partnership is the beauty and inspiring nature of the vision—the V in LVP and MVP. While the charrette drawings by artists like James Pettinari, John Koepke, and Christine Carlson are visually captivating, it is the overarching vision—the "big idea and principals/moral infrastructure," as Carlson described it—that can serve as a framework for systematically building a future for the Mesabi when the inevitable day comes that mining is over. This vision doesn't compete with active mining to achieve the landform-shaping goals that will bring about the landscape of the future. In fact, it needs and promotes mining. This vision can serve as the framework into which smaller projects can fit and make sense.

In 2023, the LVP is twenty-five years old. In comparison to its potential centuries-long timeline, it is, as Swearingen said, still young. But it has already outgrown its parents (the founders of the LVP) and now has a new generation of friends. If, in the LVP's first quarter century, it succeeded in captivating people with a vision of how mineland could be converted to a valuable environment-, community-, and job-building asset, but the timescale involved went beyond the span of human careers and the attention of its initial founders—could there be a way to create a difference on a shorter, more career-friendly schedule? And could ways be found to navigate the complex issue of when to begin repurposing minelands without sacrificing their potential for future mining? In other words, could we have some of the compelling future painted by the LVP not a century from now, but today? These are the questions I grappled with when my personal story became entangled with the history of mineland reclamation and repurposing on the Mesabi described in the next chapter.

CHAPTER 6

The Future Today

The creation of shoreland for pit lakes that may not exist until
75–100 years in the future is worthwhile but redeveloping
mine lands for near-term economic activity will have more
immediate impact for Iron Range communities.

"The Future Today: Redeveloping Mine Stockpiles &
Creating In-Pit Pit Lake Shoreland Zones," July 2013

A parade of lights glimmered off one water body after another as my
Ryder truck rumbled north through lake country along Minnesota
Highway 73. Moose Lake's thousand streetlights shone brightly off the
night-stilled waters. Cross Lake was awash in the colorful string lights
already arrayed on the docks and decks for the Fourth of July. Prairie
Lake gently strobed moonlight into the cab of my truck through the
passing trees.

My father-in-law, John, and I were covered with a salt crust of summer sweat and the dust of a garage full of tools, Weber grills, winter
tires, and miscellaneous odd endpieces of a newly emptied house.
We had packed all these dregs of a St. Paul home into the truck and
driven northward. With my young family of three, I was moving from
the heat of the city to a place where, as Frank Hibbing before me, my
bones might be chilly as I slept over nature's motherlode of iron. We
were bound for the Mesabi Range and my wife's hometown, where
we were to settle among family, friends, good public schools, and the
quiet streets of a small mining town.[1]

As I drove farther north, the lakes began to call to me. They promised to wash away the sweat and grime and perhaps give us five minutes
of stress relief as our breathing deepened and our skin tingled in the
cool water. I slowed the truck to a lumbering turn and crunched onto
the gravel drive of the public launch at Island Lake. I wondered when

had been the last time my fifty-one-year-old father-in-law had pulled over for a late-night skinny dip on a public beach. But he seemed game, and I was primed for absolution. After all, I was coming home, in a way—home to a wilder northland that exists only when you cross US Highway 2, a sort of Mason-Dixon line for many who prefer a woodsier life of hard work, tradition, and outdoor adventure. Home to a land where you can *exhale*, and then fill your lungs with clean, cool air. Home to an Iron Range of Jacobean town halls and Carnegie libraries and artisan-built houses reflecting an architecture of once-booming wealth and promise—a century old and perhaps out of repair, but still a place of substance from which one could launch into the dark, piney wilderness of the United States' northern boundary. A place my new, city-born daughter would be proud to be from. She would be a Ranger in the way that I am a Yooper—perhaps not always living on the Iron Range, but carrying in her marrow the pride of place, ingenuity, and adventurous spirit that comes from being lucky enough to live in a natural paradise. It was late June 2001.

Once settled in my new Iron Range home, I took to the woods, which is my habit whether at home or on vacation. I had imagined I would find miles and miles of trails to bike and hike, pine forests that filled my nose with the scent of sun-warmed needles, abandoned strands of public beaches to stroll, and mountains to climb—at least, midwestern mountains—just like the Marquette Iron Range where I grew up. But this paradise was somehow different. Where there were trails, they were illegal ATV trails that crisscrossed the land without any maps or official signage, punctuated with mudholes, unstructured wetland crossings, and "No Trespassing" signs that often ended at re-cently reconstructed mine fences. There were a few red and jack pines where the Boy Scouts had replanted them in the 1940s, but mostly it was a thick tangle of upstart aspen. As for abandoned beaches, you had to drive past dozens of water-filled mine pits, shining aqua green in the sun behind the fence line, on your thirty-minute drive to the public beach that charged an entry fee. And the mountains—human-made and angular, like rust-colored, odd-shaped wedding cakes—were so steep and full of loose rock that climbing them was a serious risk to your ankles, if not your life. There was the freshly paved Mesabi Trail, which linked tour location towns like a string of pearls, but it slowly dawned on me that this was not the boundless silent sports paradise

I had imagined. The pockets of beauty and outdoor intrigue felt disjointed and off-limits.

Over the next few years, my expectations were challenged in another way. With a second child on the way, my wife and I wanted to upgrade our sedan to a small SUV—something higher for snapping children into car seats and with more room to haul pack-and-plays and strollers. So we made the eighty-mile trek to Duluth to test-drive. With all the patience that a toddler and a pregnant woman can afford, we tried several dealerships. Every salesperson was our new best friend. At our final stop, we were greeted by the man in the showroom.

"Where are you from?"

"We're from the Iron Range."

"So," he said with a frown, "what did you do wrong to wind up there?"

What did you do wrong to wind up there? These words played through my head over and over in the months and years that followed. We had moved to the Iron Range—and taken our metropolitan jobs with us—for the quality of life, the family-centered atmosphere, and the space for outdoor adventure it afforded. We *chose* the Iron Range, and we created our own jobs in the process when we opened a new engineering office in town.

What did you do wrong to wind up there? I could not figure out why the salesman would risk offending a customer he knew was on the brink of handing him a good sales commission.

But as I walked the streets, talked to my neighbors, and read the local newspapers, a hypothesis dawned on me. What if the salesman was not socially inept but was only doing what good salesmen do—finding common ground? Maybe he had only reflected back to me what other Iron Rangers had been telling him: that we should not be happy to be from the Iron Range.

It was only a hunch, a string of unsettling anecdotes, until I saw the results of my new hometown's 2007 branding study. It had asked 620 community residents to assess their hometown from the perspective of whether they would recommend that someone live, conduct business, or visit there. The results were astounding. In comparison to national averages, in which respondents were ranked on how strongly their positive views of their own city were held, the Iron Range residents overwhelmingly veered to the negative. Would you recommend visiting the Iron

Range? The locals thought it was not worth the drive. How about living there? No. Is it a good place to conduct business? Definitely not. These results are republished below. The pollsters at the brand consulting company had not seen negative numbers like this before.[2]

Community Brand Barometer

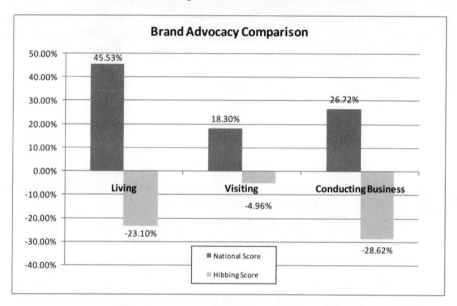

What exactly were the problems residents felt existed? The chart on page 151 shows the rankings. The residents felt that our young people did not want to stay around (so much for pride of place in my daughter's marrow!). We didn't welcome anyone new into the community, and we were afraid to change. There was no place to shop, and even if we wanted to, we did not earn enough to do it. The study reflected how residents felt in 2007.

The way we were perceived in the media didn't help. Five years before the branding study, *National Geographic* had published a feature about the town in its zip code series in which the author wrote, "Like Dylan before them, the first thing kids graduating from Hibbing High want to see is Hibbing in the rearview mirror." And about Dylan himself, they added, "his feelings about his hometown have been, at

Online Community Survey Results
(Hibbing residents' responses)

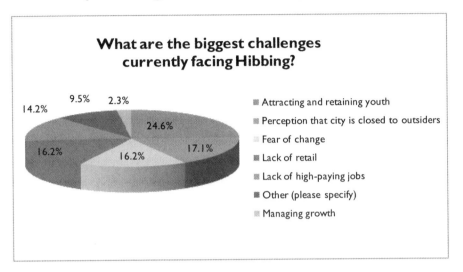

**What are the biggest challenges
currently facing Hibbing?**

14.2% 9.5% 2.3%

24.6%

16.2% 16.2% 17.1%

- ▨ Attracting and retaining youth
- ▨ Perception that city is closed to outsiders
- ▨ Fear of change
- ▨ Lack of retail
- ▨ Lack of high-paying jobs
- ▨ Other (please specify)
- ▨ Managing growth

best, ambivalent." I read the same ambivalence in Jonathan Franzen's Minnesota-based novel *Freedom*, in which "basic Iron Range squalor" is the phrase used to describe the hometown of the unlikely protagonist, an expatriated Ranger, whereas Crocus Hill in St. Paul (my old running route that I abandoned to move to Hibbing) seemed like a Fitzgeraldian utopia of mansions. While it is true that there are more mansions in Crocus Hill than in Hibbing (including the James J. Hill mansion, at least partially built with wealth from the Iron Range), to read an award-winning novelist describe your adopted homeland using the word "squalor" stings.[3]

Reflecting upon my readings and personal experiences, it seemed to me that Iron Rangers didn't love their homeland, and neither did anyone else. I struggled with the idea of living in a place that seemed to be reviled. I had lived in poor towns and wealthy ones, small and large, American and European, but never anywhere that seemed to be held in such low esteem by residents and outsiders alike.

Being fundamentally an optimist, I pledged to espouse an upbeat attitude and never say anything negative about the Iron Range in an effort to single-handedly stamp out defeatism. I organized breakfast meetups of young professionals to share the "hidden gems" of the Iron Range in order to spread pride and positivity. I did informal surveys of residents young and old to understand whether the feeling was universal—and concluded that only the oldest Iron Rangers and elected officials seemed immune to the blahs. As a rule, young professionals, children, and young parents, while they appreciated the safety and family-friendly atmosphere, all seemed to be on the verge of leaving this place. Maybe *National Geographic* was right: young people here just wanted to leave. For my generation, which only really remembered the Iron Range after the economic collapse of the 1980s, our hometowns seemed to steadily lose people, restaurants, shops, and entertaining things to do. Without a developed scene for outdoor, nonmotorized recreation to offset these losses, there wasn't much to hold people like me on the Mesabi Range. In the end, all of this talking helped me understand the malaise, but as a new resident of the Mesabi Range with virtually no professional network or connections to people in power, it seemed as though not much progress was being made to change things.

As I turned over this problem in my mind, I continued to run, hike, ski, snowshoe, and bike wherever I could on the unnatural landscape of the Mesabi. In my explorations, I came across views that were astounding: Rock stockpiles from which you could see twenty miles in every direction. Moss-covered plateaus of mined rock studded with stunted birch trees among which you could bike in any direction. Mature aspens growing through the hollows of black-rubber haul truck tires six feet across. Aside from the Mesabi Trail, the routes I explored were not sanctioned or marked; mudholes were common, as were fences and too-steep slopes, but there were pockets of curious, unnatural beauty that took me years to find and appreciate. In the book *Islands of Abandonment: Nature Rebounding in the Post-Human Landscape,* Cal Flyn describes a similar epiphany on the oil-shale minelands around her family's native Scotland: "What eyesore sites like wastelands can teach us is a new, more sophisticated way of looking at the natural environment: not in terms of the picturesque, or even the care with which it has been tended, but with an eye upon its ecological virility." She began to marvel at nature's ability to reclaim this forbidding terrain, just as I did. There was something unique and beautiful where pockets of vegetation sprouted up among the remnant artifacts from bygone industry, though

Determined growth: aspen trees reclaim disturbed lands, even growing up through a mine tire. *Author's collection*

beauty may not be the first word that jumps to mind. Flyn explains further: "To come into an abandoned mine or spoil heap or quarry or parking lot or oil terminal, and see it for the natural wonderland it has become is, I admit, a difficult task. But in these environmentally straitened days, it is a taste worth cultivating."[4]

She goes on to invoke the French term *jolie laide*—"literally, 'pretty-ugly,' a term applied to women whose imperfections elevate them from conventional attractiveness to a higher plane of visual interest"—and note how "the bings [Scottish term for oil shale waste dumps], and places like them, might be a jolie-laide landscape: ones whose industrial scars only serve to throw their current stature and ecological significance into sharp relief."

I began to wonder how the critics of the Iron Range would feel if they could somehow see and experience these small, unusual wonders. And if something could be done to enhance them, promote them, and provide legal access to them. Might such glimpses change people's minds? After all, evidence of community pride restored had been documented as the result of derelict minelands being reclaimed elsewhere in the world. And if that goal was too lofty, at least such an effort might give us something to do close to home (like mountain bike or hike) and maybe keep or attract young professionals and families to our area. I imagined that reclaiming the mining landscape could be like a house-painting project, restoring the beauty of the underlying structure—in our case a foundation of good education, family-friendly community, reasonable cost of living, and proximity to the outdoors—while giving us a reason to stand back at the curb and give a self-satisfied nod.[5]

My explorations occurred indoors as well as outdoors. At the public library, I soaked up the history of the landscape I trod, reading *Seven Iron Men*, Marvin Lamppa's *Minnesota's Iron Country*, and Edmund Longyear's *Mesabi Pioneer*. One day, I perused the local reference section, where things such as city meeting minutes and master plans were kept. That was where I came across a white three-ring binder labeled "Laurentian Vision Charrette." In it, I saw that someone else had dreamed my same dream. The charrette drawing illustrated the pines of Boy Scout Hill that I had grown to love enshrined as a regional park, with a tree-ensconced creek flowing south from one greenway to the next in a vegetated corridor. The closed mine pits were shown as waterfront developments surrounded by a forest full of trails. It was

The Hibbing charrette drawing (2003) included Boy Scout Hill Regional Park, which was proposed to be a green space ecological corridor running south to north across unreclaimed minelands. *Courtesy of Northwest Architectural Archives, University of Minnesota Libraries*

just a glimpse—I saw it only once, and I cannot recall exactly when—but it was an inspiration. At the time, I had no idea where it had come from or who had made it.[6]

In 2008, amid the community malaise captured in the branding study, it seemed as if mining was poised to surge into new technology and investments that might breathe life into the Iron Range. Magnetation had just opened its first plant that employed a proprietary magnetic separation technology to recover iron from tailings and lean ore stockpiles left behind by John C. Greenway and other conservationists of our iron ore resources. And India-based Essar Steel was on the brink of a $1.65 billion investment to build the Iron Range's first steelmaking plant, thereby eliminating both the step of reheating the iron (one of the major costs of both taconite pelletizing and steelmaking) and the need to ship our iron ore off to the mighty blast furnaces of the lower Great Lakes. These innovations were expected to bring more jobs back to mining than had been seen since before the collapse of the 1980s. There was talk of growth and the need to assimilate the wave of workers who would repopulate the towns of the Iron Range. At one regional economic conference I attended, this hopeful attitude was given a soundtrack when the audience was warmed up by the 1980s one-off hit "The Future's So Bright, I Gotta Wear Shades."

The optimism about a possible resurgence in mining seemed at odds with the futurelessness the people of the Iron Range expressed in the 2007 branding study. Later, I came to attribute this community listlessness to a sort of economic development fatigue—the sense that the next big thing never seems to happen. Rangers have their own term for this economic salvation that never arrives: a "chopsticks factory," a reference to an ill-fated attempt in the 1980s to combat the perceived loss of US steelmaking jobs caused by increased imports of Asian cars by selling up to 6 million pairs of Iron Range–made chopsticks back into East Asian marketplaces.

■ ■ ■

And so it was, sporting optimism tinged with doubts about another chopsticks factory, that over a thousand Iron Rangers, plus Minnesota governor Tim Pawlenty, gathered on a sunlit grassy field north of Nashwauk. We were there to break ground—and, in a good-luck tradition from India (the home country of the prospective mine developer, Essar Steel), to break coconuts—to usher in the new mine and steelmaking plant and the next wave of Iron Range economic prosperity. Among the people in attendance that day was Jim Plummer of the Iron Range Resources and Rehabilitation (IRRR) Board, who happened to stand next to me as the dignitaries posed for photos and prepared their shovels.

Plummer was a sandy-haired, ebullient man in his forties. It was clear he believed in fitness and the life-changing possibilities of seren-dipitous meetings. Having grown up in suburban Burnsville, Minne-sota, but with roots in Iron Range mining, he loved the piney patches of the north. When I met him on that grassy field in Nashwauk, he was fairly new in his role as the IRRR's coordinator of the Laurentian Vision Partnership (LVP). As we got to talking, he became the first recipient of my mind dump of ideas, developed thus far in isolation, about how we could transform the Iron Range, and perhaps people's attitudes about it, by repurposing the inactive parcels of our mining landscape. My idea was not repurposing instead of mining; it was to pursue simultaneous repurposing *and* mining, in a way that would use the land and mined landscape to the best effect. The key was to do it *now*, I explained, not centuries into the future when mining was over and the towns of the Iron Range may have further declined or alto-gether disappeared.

In retrospect, my ideas seem sheepishly unoriginal. I was gushing about ideas that people like Sam Dickinson, Dan Jordan, Jim Swearin-gen, Dennis Hendricks, John Koepke, and Christine Carlson had been preaching for decades. But at the time, it felt revolutionary. Sensing my raw enthusiasm, Plummer invited me to a meeting of the LVP to share my ideas. It seemed as if I had gained a partner, and as he grasped my hand with a firm grip and flashed an easy smile, he left me with the words, "I'm voting for you for mayor."[7]

By November 18, I had prepared my first presentation to LVP: "From Red to Green: Planting Hibbing's Mine-Affected Proper-ties." It espoused the forethought of the Boy Scouts, whose plantings in the 1940s now softened the oblique angles of the red ore dumps that flanked Hibbing's eastern border with a peaked green canopy of sixty-foot-high pines. Because of the pines, I argued, Boy Scout Hill had been repurposed from a mere rock dump into a favorite place for four-wheeling and campfires and the site of a large multifamily apartment complex known as the Southview Apartments (named for their southern view off the dump). The proposal was simply to con-tinue what the Boy Scouts had started by planting on the most visible mine-affected areas around town, thereby adding value to *any* future use of the property—timber value, if it were someday remined, or aes-thetic value, if it remained untouched by mining—and enhancing the

Boy Scout Hill in Hibbing, its trees planted on red rock dumps in the 1940s, was a model for the early reclamation concept that later developed into a mountain bike park. *Author's collection*

overall community image. What the presentation lacked in terms of originality or serious technical merit I made up for with cute images of my children, Francie (age seven) and Oren (age five), who were pictured scrambling up the needle-strewn rocks of Boy Scout Hill. The presentation received a tolerant acceptance from the LVP members and, for all its flaws, was successful in getting me an audience with the Minnesota Department of Natural Resources (MDNR) Lands and Minerals staff in January 2009.

■ ■ ■

The temperature was –23 degrees Fahrenheit on the morning I set out for the meeting with MDNR officials at the Hibbing Lands and Minerals office. The staff provided me with a map of the unreclaimed mine stockpiles and landownership in Hibbing, which I now know to be the atlas created by the LVP. They advised me that the stockpiles might stand the best chance of growing if planted with jack pine, mixed with red pine and cracked corn. They said I wasn't the first citizen to recognize the need to plant the pre-1980 unreclaimed mine sites around town: former Minnesota governor Rudy Perpich had planted an area west of town a decade or two earlier. The meeting gave me enough to stay interested.

Armed with the maps provided by the MDNR, throughout the spring and summer of 2009 I performed reconnaissance of the barren stockpiles most visible from the middle of downtown Howard Street. Oblivious to the past, I was inherently drawn to the idea of "visual screening" that had been so hotly debated during the development of the Mineland Reclamation Rules. But I was looking at pre-1980 stockpiles that had had no revegetation requirements during the natural ore mining period. As noted in chapter 1, Hibbing is literally surrounded by barren pre-1980 mine dumps that are visible from the center of town; therefore, the challenge was not finding a suitable, unvegetated stockpile but instead deciding which one.

Given the rough but roadless terrain, my vehicle of choice for reconnaissance was my 1998 Trek mountain bike. It had front suspension to absorb the shock from the rocks, knobby tires for grip, and twenty-one gears for climbing the steep sides of the mine dumps. I had started mountain biking in the Upper Peninsula of Michigan in 1989 when my parents bought me a rigid-framed Schwinn High Plains for Christmas. But while living in St. Paul, I had felt compelled to upgrade after riding the state's first dedicated, purpose-built, single-track mountain bike trail at Lebanon Hills in Eagan. In 1998, with leftover per diem from work travels, I had bought a pair of new Treks for my wife, Miriam, and me.

The bike proved ideal for the mine rock terrain, enabling me to get around berms and lift, then hop, over fences while covering ground at speeds unattainable by hiking. Riding the old mine haul roads and illegal ATV trails, I made my way around the potential project sites. And then my simple tree-planting idea began to transform. I began to envision terraforming and purpose-built, natural-surface, single-track mountain bike trails like the experimental one I had ridden at Lebanon Hills.

With this headful of fresh ideas, I invited Jim Plummer for a tour of the potential project sites. I had chosen the overburden pile of the Gray Reserve west of Hibbing, a huge mound of sand and gravel being made incrementally smaller every day by an aggregate operation that was chipping away at its toe. There was also the Susquehanna lean ore stockpile east of town, perhaps Hibbing's most visible mine dump, which the members of the cross-country team had dubbed "the Top of the World" as they ran up the cobble- and beer can–strewn haul road to the top. And there was another, shaley rock stockpile from the Susquehanna Mine, this one north of town inside the former Boeing

In 2008, the author and Jim Plummer rode on the Susquehanna Mine rock dump north of Hibbing, an early contender for what would become "a world-class" mountain bike park. *Author's collection*

Mine and overlooking the five-hundred-foot human-made cliff down to the bottom of the Hull-Rust open pit. The common elements of these sites were their proximity to downtown, their dramatic views, and their well-drained, compacted surfaces that appeared ideal for trail construction. At the time, these dumps seemed like an obvious place to try building some trails—and in fact, at Cuyuna, a hundred miles to the southwest, trail builders would later discover that the stockpiles left over from iron mining were almost perfectly suited for the construction of naturally surfaced trails. (Typically, the first thing trail builders do is remove the soft, organic topsoil, but on the sparsely covered pre-1980 stockpiles, the topsoil was still thin or nonexistent from having been disturbed by mining and the ground surface was made of compactable, well-draining mine spoils.) Although the site we selected that day was about three miles from its eventual home and was not yet named, our excursion on July 31, 2009, was the first semi-official ride on what *Outside* magazine would hail twelve years later as "a world-class mountain bike trail system."[8]

■ ■ ■

After the success of the first test ride, I officially pitched the concept of a mountain bike park to Plummer and his boss, Dan Jordan, who memorably called it a "sexy idea." But what followed for the next two years was a remarkably unsexy bureaucratic slog of research, board presentations, one-on-one sales pitches, question-and-answer sessions, and phone calls to float the idea. My calendar shows that I took about thirteen half days of vacation to give about fifteen different pitches as my third child, Aida, grew from newborn to toddler. The common reaction to the pitches, as I recall, was a room full of silence. By this time, mountain biking had transformed some mining towns, like Moab, Utah, into outdoor recreation meccas. But in Minnesota, Cuyuna's trails had not yet opened, and the Cyclists of Gitchee Gumee Shores (COGGS) were just getting started on trail development and management in Duluth. A mountain bike park on the Mesabi Range, to many Iron Rangers, smacked of chopsticks. But after each speech, there was often a single person who approached me to express intrigue at the possibilities. At one Rotary Club presentation, trying to read the tenor of the silence in the room, I handed out note cards and asked each attendee to write down whether they supported my idea or not.

I will never forget the eighty-year-old woman who—though she professed to not be a mountain biker—handed me the card faceup with the words "100% supportive" scrawled across the front. One by one, I was winning people over.

The project received some critical momentum in 2011 in the form of a shaven-headed, brash, and serious young landscape architect named Jim Shoberg, who hailed from Duluth but was originally from Chisholm. At first, I had no idea what his intentions were. It seemed to me that he just appointed himself to the team and began contributing ideas that started with the pronoun "we." Finnish Americans are famous for our attitude of self-reliance, sometimes to a fault—after all, Helsinki's fort at Suomenlinna is inscribed with the phrase "do not rely on the help of strangers." But in May, I decided to swallow my reservations and get to know Shoberg on a tour of the North Hibbing project site. It was immediately obvious that he was passionate about mountain biking and mineland reclamation. He knew the players in the emerging northeast Minnesota mountain biking scene, such as Hansi Johnson and Aaron Rogers, both with the International Mountain Bicycling Association (IMBA). He understood landscape design, and he had been successful in getting grant funding for mine reclamation projects. That day, we joined forces. For me, it was a lesson that would prove true at least five more times as the project played out: when well-intentioned and passionate people who share in the vision come forward, the project and the team should flex to welcome them.

Shoberg, Plummer, and I planned to build the mountain bike park using money from LVP's recently established innovation grant program, which at that time was funded at $250,000 per year. The mountain bike trail qualified for the grant because it featured a form of mine reclamation or repurposing that went above and beyond the requirements of the Mineland Reclamation Rules. The grant funds, along with the rest of the IRRR's budget, were provided through production taxes on taconite mining. Because the stockpiles and mine areas of our project site had been created before 1980, there were no rule requirements at all, so any type of reclamation work would qualify. We also targeted federal transportation funds. But in order to be eligible to receive the funds, we needed a public partner and fiscal agent. Thus, in 2011, we began discussions with the city of Hibbing.

Working up from the parks and recreational board, through the director, and on to the city council, we presented the project and its merits. As before, the presentations received a mixed reaction, but we typically won one devotee at every level. Some of our new project supporters had been active in the LVP, understood the concept of mineland repurposing, and had championed some past LVP innovation grant projects. It proved timely that the mountain bike trails at the Cuyuna Country State Recreation Area had opened earlier that summer. On September 13, 2011, a caravan of city officials and staff from Hibbing drove down with our project team to see how the Cuyuna trails had been woven into the aspen and pine forest that had sprung up on the postmining landscape. The trails were a tangible demonstration of what we were trying to do on the Mesabi, with the principal difference being that Cuyuna's mining had stopped decades ago, whereas taconite mining on the Mesabi was still vital. After this visit, the city officials understood that we were not proposing another twelve-foot-wide, paved Mesabi Trail passing through the community. In contrast, our project would feature a two- to four-foot-wide natural (dirt and rock) trail, swooping up the slopes and between trees, with a single point of access and a parking lot. If we built it well enough, we argued, the trail would bring people to town and keep them there, not merely bring them *through* town on their way to another destination. In this way, it might support and grow our local economy. Yet in 2011, it was still too early to see firsthand the boom in business creation and job growth that Crosby, Ironton, and Cuyuna would experience over the decade to come. At least we could point out the number of red-tired bikes atop bike racks (the dirt's hue providing a clear indicator of what had attracted the biker to town), hear local business owners' stories of increasing sales, and view the sign that Dairy Queen had voluntarily erected showing where patrons could ride mountain bikes.

■ ■ ■

For the next year and a half, Shoberg, Plummer, and I worked to solidify the concept. Shoberg brought in Aaron Rogers and Hansi Johnson to help us explore the trail layout. From February to August 2012, we developed seven different design iterations, ultimately settling on a concept that included fourteen miles of single-track trail, two parking lots, and improvements to North Hibbing's existing campground. Aside from the IRRR's donation of Plummer's staff time, it was an all-volunteer effort. I worked more than 150 volunteer hours in 2011, equivalent to nearly four weeks of vacation time, which Shoberg and the others surely matched. But our efforts were bearing fruit. By mid-2012, we had a formal agreement that the city would act as fiscal agent and the administrator of a "conservation park" (not a city-maintained park, but a natural area), land agreements in process, and over $1.1 million in grant applications awaiting consideration.

Another important development occurred in late 2012. To this point, the project had been known as the Susquehanna Mountain Bike Park, after the nearby Susquehanna Mine, the point of origin for the natural ore rock stockpile upon which the trail was to be built. In project correspondence, Minnesotans nearly always misspelled Susquehanna, including on the underlying MDNR map. It seemed like a mouthful, much like Cuyuna before people got accustomed to it (I once heard an elected official accidentally mispronounce that name as *cojones*). Furthermore, it was an eastern name, hailing from Pennsylvania, that did nothing to promote our region's sense of place. (Many mines around Hibbing are named for eastern locales, such as the Albany, Scranton, and Mahoning.) Instead, I wanted a memorable name that both reflected our region and implied a sense of adventure. In this spirit, the name Redhead Mountain Bike Park was born: "red" to salute our red hematite ore, and "head" to imply the beginning of an adventure and a place of recreation, like trailhead, Palisade Head, Hilton Head, and so on. Our region already existed at the head of many things, including North America's principal watersheds and its supply of raw materials for the steelmaking industry. This colorful name, we hoped, might honor our unique position in the American landscape and inspire us to adventure in our own backyards. Plus, it rhymed and was easy to spell.

As we planned to have the grand opening of the park in June 2013, however, our beginner's luck ran out. Unwittingly, we were headed full steam into the period I would later think of as Redhead's first death. It was to be a death by starvation.

With a development plan in place, the first $250,000 in grants secured, and a draft land agreement on the brink of being finalized, calls to our municipal partner began to go unanswered and emails unreturned. There were new people in office with a different set of priorities. The city dissolved its economic development functions and no longer participated in the LVP. The change in attitude came to a head on April 19, 2013. In an effort to get the city reengaged, Plummer and I held a meeting with the mayor, a newly elected city councilor, and the provost of the local community college, Mike Raich, whom we had recruited as a potential project supporter and maintenance partner for the trails. The city's elected officials came into the meeting cautiously and defensively. At one point, when asked what the city could do to support the project, the mayor shouted angrily, stood up, and walked out. It was so awkward and unusual that Raich called me later that day to ask whether I was okay and if city meetings usually ended that way. I do not know whether the city's reversal was prompted by resistance to the idea of repurposing minelands (which had received some criticism in a letter to the editor of the *Hibbing Tribune*) or by a financial inability to embrace a risky new idea when city resources were slowly draining away decade after decade. In any case, it seemed Hibbing was no longer the razzle-dazzle town it had once been, yet, being politically inexperienced, I didn't know that we should have dissolved the partnership with the city at that time. Instead, we tried—in vain—to make amends with the city and see the project through. Slow action turned to inaction, and the land deal eventually faltered. Jim Plummer and I decided, under the advice of IRRR commissioner and project supporter Tony Sertich, to simply let the grant expire at the end of 2014, at which point the money would return to the IRRR coffers due to nonexecution. There was nothing to do but let the project die.[9]

Seth Godin's book *The Dip* describes the point in a new venture when your initial momentum is lost and it feels like you are failing. You need to decide whether to quit or to continue. We were at that point for Redhead.

Recollecting that time calls to mind the Finnish word *sisu*, which has no direct English translation but whose meaning is still understood on the Finnish American–influenced Mesabi Range. The BBC once attempted to define sisu as strength, resilience, and "perseverance in a task that for some may seem crazy to undertake, almost hopeless." Crazy and almost hopeless. The Redhead project at this point had no city willing to support it, no land, and soon no money. However, in the first six years of pursuing this vision for a mountain bike park in the middle of the Mesabi Range minelands, we had developed several things. We had a mountain bike club—Iron Range Off-Road Cyclists

First club ride, Iron Range Off-Road Cyclists (IROC), 2013. Club members volunteered to build, ride, and maintain single-track trails on the Iron Range. *Author's collection*

(IROC)—a small, loose-knit group of dedicated volunteers who had coalesced around the possibility of Redhead and had hand-built miles of single-track trails into the nooks and crannies of the range, including Hibbing's Maple Hill, Virginia's Lookout Mountain, and various dumps and stockpiles. We had the full support of the commissioner of the IRRR. We had a few converted individuals scattered around the communities who had probably softened to the idea of repurposing the minescape from the decades of work completed by the LVP. And most importantly then and still today, we had what Dan Jordan had called the "sexy idea" that mountain biking could transform our community in a way that had proven itself in places like Moab and was beginning to prove itself in our sister range to the south, the Cuyuna. And finally, we had a killer name. The question was, could these assets that had coalesced around the first potential site for a mountain bike park in Hibbing be lifted and placed elsewhere—perhaps even somewhere that was *better* suited to the idea?[10]

■ ■ ■

In late March 2014, I began seriously considering alternative sites for Redhead. In engineer fashion, I created a spreadsheet with a ranked matrix of desirable site characteristics, which included 1) that it must be mineland, 2) that it should be close to a population center, 3) that it would have the significant topographic relief and area needed for mountain biking, 4) that it be close to or, ideally, adjoining the Mesabi Trail and other recreational amenities, and, importantly, 5) that it be predominantly publicly owned, to remove the variable of having to engage a private party in a land deal. In some ways, it was almost unnecessary to have listed the first item because the *only* sites near our towns with large enough undeveloped areas and significant topographic relief adjacent to the Mesabi Trail were minelands. Everywhere else was either too far from town, too small, too flat (and, therefore, wet), or too distant from a cluster of other recreational amenities that would help the nascent mountain bike park to succeed.

By this time, it was clear the huge open pits and stockpiles could provide the ideal topography for mountain biking. As mentioned in previous chapters, the difference in elevation from the bottom of an open pit—even one flooded with groundwater—to the top of an adjacent stockpile could easily be three to five hundred feet. The terrain is undulating, which can create the roller-coaster experience critical to the new type of purpose-built mountain biking trails. And the bare bedrock, crushed-rock stockpiles, and overburden piles provided the near-perfect surface for trails. IMBA had told us that in other parts of the country, like the clayey southeastern United States and nearby Duluth, preplanned mountain biking events such as the Red Bull Rampage would often turn into dangerous mud fests after an overnight rain. But we had learned that our building material was almost tailor-made to avoid that outcome. In Cuyuna, which had similar mine-created materials underfoot, the trail builders had taken to calling it "red gold." The more I learned about the unique properties of the minelands for natural-surface trail construction, the more I thought of it as a natural recreational advantage for our area—akin to how the positioning of the Great Salt Lake and the Wasatch Mountains leads to the lake-effect snowfall that makes Park City, Utah (another former mining town), such a great ski area, or how the geomorphology of Florida makes for great beaches.

Importantly, the naturally surfaced trails we were proposing would leave a light and nonpermanent footprint on the minelands. Unlike the Cuyuna Range, where all mining had stopped decades earlier, mining on the Mesabi was still a thriving economic venture that underpinned our economy. Part of the reluctance to embrace wholesale repurposing of the idled mine pits and stockpiles was the fact that they still contained iron units—carefully set aside at cost to the mining companies by miners such as John C. Greenway, who adopted the conservation principles of Theodore Roosevelt and Gifford Pinchot. But in comparison to heavier-footprint undertakings, such Nashwauk's controversial housing development—which required not only houses but roads, natural gas pipelines, sewer, and water services (in other words, things that could not easily be moved once installed)—trails were truly nothing more than cleared paths cut into the side of the pit or a stockpile. If economic conditions favored remining the area of the proposed mountain bike park, the trail might easily be relocated, perhaps to a spot that had just completed a cycle of mining. We considered the trails to be a "temporary" or "intermediate" use of land that might someday be returned to mining—a philosophy that would provide the most benefit to the most people rather than leave the land idle between periods of active mining, as it had been for decades. In this way, trail recreation and mining could be symbiotic, with mining creating the dramatic landscape and great substrate for trail construction and the trails providing the quality-of-life amenities that would attract and retain young professionals to the mines and businesses of the Iron Range and diversify our economy through tourism. It was simply a new form of the multiple resource management philosophy embraced by Sam Dickinson's Erie Mining Company.

■ ■ ■

The need to be close to town and other recreational amenities was a lesson already being learned at Cuyuna. When it opened in 2011, very little infrastructure to support the trails had been contemplated. There was just a small dirt parking lot, a handful of other recreational offerings, and a few local restaurants. But after three years in operation, its popularity had grown to the point that the dirt lot was getting overwhelmed on weekends and the area needed more restaurants, overnight accommodations, and other recreational attractions to serve the new demographic of people visiting the town. Having experienced this crowding personally, I hoped to locate our mountain bike park near a town that already had other recreational offerings, restaurants, gas stations, and hotels that might serve, and be served by, the mountain bike park. And, of course, we needed a new municipal partner with a more optimistic outlook toward this promising if little-proven idea.

In April 2014, we began to explore the idea of positioning the mountain bike park in Chisholm, on lands predominantly owned by the IRRR and near the Minnesota Discovery Center (previously known as Ironworld), a museum of the Iron Range that came ready-made with a campus containing a huge parking lot, restrooms, a restaurant, an outdoor amphitheater, and a paved trail spur connecting it to the Mesabi Trail. With Plummer's, Jordan's, and Sertich's stalwart support, the IRRR was a shoo-in, so I began to focus on the other potential partners. I had three meetings each with representatives of the city of Chisholm and the Minnesota Discovery Center. These were just quiet meetings—exploratory affairs—to determine whether further pursuit had any merit. I recall one conversation with city administrator Mark Casey in which he proclaimed, "In Chisholm, we like to think of ourselves as the City of Yes." With enthusiasm building, I sent Jim Plummer another appointment for a bike-borne site exploration, this one cryptically entitled "Bike around Chisholm," as we were still formally betrothed to our original site until the grant money expired.

On a June day, we used our mountain bikes to descend from the parking lot of the Minnesota Discovery Center via the old mine haul road, past the gates and fences surrounding the mine pits, and into a complex of ten former natural iron ore mine pits. It was a three-hundred-foot descent from the pit rim down to the aqua-green waters that had refilled the mine pit complex. The pits included the Glen, the Monroe, the Tener, the Godfrey, and portions of the Pillsbury Mine that had originally been dug around 1900. Underground mines were present as well, though they had mostly collapsed or been mined out by the open pits. The areas had been operated through the mid- to late 1960s, the final ore shipments made roughly thirty-five to forty years before our reconnaissance.

As we sped down the haul road on our bikes, we saw a minescape in transition back to nature. The mine complex had come to resemble a red rock canyon—like someplace you might hike in Sedona, Arizona—but bottoming out in an aquamarine, groundwater-fed pit lake. Some parts were raw—steeply eroded slopes on the pit wall that resembled the tawny spires of Bryce Canyon. Other areas seemed surprisingly natural, like the spring-fed creek that bubbled up on the pit's western edge and meandered through a birch and aspen forest and vertical canyon walls down to its terminus at a twenty-foot waterfall. There were groves of red and jack pines. And there were the remnants of rail lines that had been traversed by the dinkeys filled with iron ore. There were abandoned tailings pipelines that had carried off the wastes from the wash plants and even utility poles with telephone lines and copper wires still strung or dangling like vines down to the ground amidst the overgrowth. It was a scene steeped in industrial history juxtaposed with natural repopulation that few could imagine would inhabit the bottom of an open mine pit, except for perhaps Gilbert Leisman, Sam Dickinson, or Cal Flyn. It seemed like the perfect spot for Redhead. But we could take no formal action until our first grant expired, so we would need to wait until 2015.

The minescape offered red rock canyon views, a "Bryce Canyon" scene, and spring-fed creeks and birch forests. *Top, courtesy of Joe Trevelan; others, author's collection*

On January 22, 2015, the first official meeting with the city of Chisholm regarding Redhead took place, attended by me, Jim Plummer, Dan Jordan, Mark Casey, and Steve Cook, a former city councilor and now an officer of IROC. We began to rebuild the concept with Chisholm as the trail sponsor. The trail would be constructed in two phases located on 650 acres of IRRR-owned and tax-forfeited properties in and around the Monroe Pit complex. Phase 1A was to be a 5.25-mile trail descending into the pit from the north along the former mine haul road and crossing an isthmus that separated the water-filled Glen and Pillsbury Pits. Phase 1B was to be a 3-mile trail completing a loop around the pit complex using part of the Mesabi Trail. Phase 2 was to be 10 to 15 additional miles of infill and expanded trail on public lands. The goal was to have 20 to 25 miles of trails, which we had learned from IMBA was the threshold at which a trail crossed from being a "local" amenity to become a "destination" attraction—in other words, something you would be willing to drive a few hours to get to and that would entertain a rider for more than one day.

The process of refining the Redhead concept with the new project partners and the new site took the whole next year. Liability for the landowners and the city was a concern. The city's attorney got involved. Research on Minnesota's recreational immunity statutes showed that they protected landowners so long as trail access was free to the public and trail operators were not negligent in the design, operation, or maintenance of the park. Inquiries to the League of Minnesota Cities confirmed that the city of Chisholm's insurance would cover park operations for an increase of less than $100 per year. A risk management plan for Redhead included signage, inspection, and maintenance per IMBA's best practices. We worked through many versions of project maps and at least five iterations of cost estimates and grant applications. Steve Cook engaged the community of Chisholm and secured letters of support from the chamber of commerce, the economic development authority, the public school district, the city's parks and recreation director, the local Kiwanis club, the Kids Plus after-school program, the St. Louis County commissioner for the district, the ambulance service, and several local businesses. The Minnesota Discovery Center wrote a letter of support and committed to operating the park's trailhead facilities, including use of the center's restrooms and parking lot. All of this work was done with no grant money and at no cost. It was a labor of love and commitment to the vision by the members of the community, the city of Chisholm, the IRRR, the Minnesota Discovery Center, and IROC.

Seven years into the project, Redhead was better than it had ever been. The site was more dramatic, with a deep red-rock canyon more like something you would see in Utah than Minnesota. The landownership was more straightforward. The trailhead amenities were more attractive. The partnership was stronger and better resourced, with committed city leadership (led by Mayor Mike Jugovich) and staff, the IRRR fully engaged, the Minnesota Discovery Center willing to play the important role of trailhead host, and IROC poised to provide a small platoon of volunteer trail maintainers. My new position at Barr Engineering Co. gave me access to better mapping resources, cost-estimating tools, and experts in disciplines from surveying to landscape architecture. The citizenry and community organizations were more engaged, including one enterprising individual who erected a sign in Chisholm proclaiming it the Mountain Biking Capital of the World before any trails had even been built. There was a mix of professionals and volunteers, public and private parties, state and local governments, and nonprofit organizations all pulling toward the same goal. In the book *Our Towns*, James Fallows and Deborah Fallows, who flew across the country studying small cities, cited the importance of genuine, strong public-private partnerships as a hallmark of communities that were successfully making the transition to better fit into the new American economy. Something else they observed to be a feature of every successful small community? Trails.

From what felt like a strong foundation, I provided a project update to the LVP coordinating committee, which would be asked to approve a new $250,000 innovation grant for the project. A committee member, also an MDNR staffer, said, "Well, that's an interesting project, but it's illegal." He specified that Redhead would violate the Minnesota Statute 180 provision that required all open mine pits to be fenced off from the public or provided with some other means of safeguarding the public from the dangers of an open pit. All over the Mesabi Range, he said, landowners regularly received notices of violation for openings in their fences (most often created when members of the public cut them illegally to gain recreational access to the open pits), which they had to repair. How could we propose a project that *purposely* created openings in the mine pit fencing and invited the public to recreate on the mined land? We argued that the St. Louis County mine inspectors had given their consent to the project. He said it didn't matter; they did not have the legal mechanisms to approve the breaches in the fence. But, we objected, what about Cuyuna, which at this point had been operating under administration by the MDNR for four years on minelands subject to the same statute? "Cuyuna is illegal too," was his response. How about the inactive mine pits that had refilled with good groundwater and that the IRRR had for decades stocked with trout for recreational fishing? Also possibly illegal.

Iron Rangers had been recreating in the mined landscape for decades, like John Dougherty and his friends who had built their ski jump in the 1940s. None of the LVP charrettes had been concerned about the technicalities of building below the pit rim. People had been fishing in the stocked mine pits for decades without issue. And Cuyuna was doing exactly the same activity under the direction of the MDNR division of Parks and Trails.

But he was right: Cuyuna *had* been operating illegally. Because there were no longer any active mines, Crow Wing County did not have a designated mine inspector's office responsible for enforcing Minnesota Statute 180. I remember telling Aaron Hautala, president of Cuyuna Lakes Mountain Bike Crew and public figurehead for the burgeoning mountain biking scene in Crosby and Ironton, that his park was illegal. "Good," he replied, ever the marketer, occupying a world where there is no such thing as bad PR. But his face was grim, and he stared off into space as he took in this information.

The public boat launches on the fish-stocked mine pits were, it turned out, in a legal gray zone. They had been sloped for safe access and sometimes sleeved with fence line down to the water's edge, but there were possible loopholes and unanswerable questions about measures of safety once a fisher was on the water.

We decided the best course of action for the project to proceed was to change the state law regarding mine pit fencing. In consultation with the St. Louis County mine inspectors, the land commissioner, the IRRR, and lobbyists—who remained supportive of Redhead despite the legal hurdles ahead—Minnesota Statute 180 was redrafted to permit an exemption to the mine pit fencing requirement for government-sanctioned recreation areas. The provision would not only enable the construction of Redhead but also create a retroactive legal status for Cuyuna's mountain biking, as well as fishing, swimming, and scuba diving in the mine pit lakes.

Our local legislators, Senator David Tomassoni of Chisholm and Representative Jason Metsa of Virginia—sponsored the statutory changes. They would allow exemptions to Statute 180 if:

- It was property owned, leased, or administered by the IRRR;
- It was used for construction, operation, maintenance or administration of 1) grant-in-aid trails, 2) municipality-owned or -leased parks or open areas for recreational purposes, including naturally surfaced trails or paths, or 3) recreational use administered by a municipality;
- It was being used for economic development purposes; or
- The mine inspector determined the area was provided with barriers/signage reasonably similar to standards or that the area did not constitute a safety hazard.[11]

The changes were introduced in the 2015–16 legislative session. While a risk management strategy had been carefully constructed for Redhead, questions about the statewide implementation of this change still loomed. In the final hours of the legislative session, the law enabling the exemption failed. Redhead was dead again.

■ ■ ■

If Redhead's first death was starvation, its second death felt to me like a stabbing in the back. My years of volunteer commitment to the project, with nothing concrete to show for my efforts, were beginning to take their toll at home. While I had been dedicating hundreds of volunteer hours to Redhead and IROC, Miriam was being overburdened with an unequal share of the homemaking duties. As I reckoned with the personal cost of Redhead, I scrawled in a journal, "Stop pushing the bike park. If people ask me to work on something *productive*, say yes. Otherwise, let it go."

Despite my personal misgivings, I remained the vice president of IROC. During the next few years, president Luther Kemp and I tried to keep the mountain bike club, which would be a critical component to the operations of Redhead, enthused and growing. We wrote grants for non-Redhead projects (started by volunteers even before IROC became a formal club) so our trail bosses could construct new sections of trail in Hibbing and Virginia. We hosted the IMBA Trail Care Crew to teach our volunteers how to construct and maintain single-track, natural-surface trails. We invited Hansi Johnson and former Duluth mayor Don Ness to the Iron Range to speak to our membership about how mountain biking had played a pivotal role in that city's transition from a declining industrial city to a vibrant "city on a hill" some likened to Minnesota's San Francisco.

In 2016 and 2017, meetings about the fencing law were held with elected officials and representatives of the mining industry, the MDNR, the IRRR, the city of Chisholm, and St. Louis County. The proposed change in state law stimulated conversations about what types of use were safe for mined lands. Mining landowners, accustomed to complying with the fencing law, questioned the wisdom of an exemption that would invite the public into an inactive mine pit to perform a possibly dangerous recreational activity. Dan Jordan remembered, "Some of those conversations were *pretty interesting*." But he argued that you could go to any public beach and see a sign stating, "No Lifeguard on Duty—Swim at Your Own Risk." He asked, "If it has worked at beaches for years, why wouldn't it work here?" Liability and trespass seemed to be the main issues for landowners who held property on the iron formation.

In 2017, I spoke to the LVP (January), the Minnesota conference of the Society for Mining, Metallurgy, and Exploration (April), and the national Transportation Research Board conference in Duluth (July) about the possibility of *temporarily* using minelands for recreation as a way to boost the mining industry's social license to operate. The idea was that—provided these liability and safety questions could be addressed—mineland repurposing need not wait until all future mining potential was exhausted and everyone agreed that mining was over. Given the economic and geological conditions on the Mesabi Range, that day would likely be decades in the future, if it ever came at all. In the meantime, the mining industry was undergoing heavy opposition based on claims that it rendered land unusable that conflicted with the ability for other industries, such as tourism, to flourish. Why not counter these claims by demonstrating that minelands could be put to productive uses even while they were awaiting the technology or economic conditions that would make them profitable to mine? Recreation was both highly visible and accessible to the public—something everyone could experience for themselves. Northeast Minnesota already had a well-developed recreation sector into which the repurposed minelands would nicely fit. The question was, Might it not change the minds of some people who were opposed to mining if they had a fun and healthy experience riding a mountain bike inside an inactive mine pit instead of just seeing it as a hole in the ground walled off by fences and "No Trespassing" signs?

While this concept piqued interest, there were some legitimate concerns that any investment in recreational development that was understood to be "temporary" was a waste of money. Why build anything that you might need to tear down someday? Now the economic impact studies that were beginning to emerge from other recreational destinations, like Cuyuna, became helpful. For example, if, like Cuyuna, a recreational asset was estimated to pump $2 million per year into its host economy and it only cost $1.8 million to build, the return-on-investment period was less than one year. In Minnesota's regulatory environment, the period for environmental review and permitting of a new mining operation is typically measured in multiple years or decades. Therefore, even if a site was proposed for future mining, every year of intermediate recreational use after the return-on-investment period was gravy for the local economy.

Another legitimate concern was over the social blowback that might be suffered if a mineland that had been dedicated to recreational use was subsequently returned to active mining. In other words, would it be worse to let the public love and then lose a recreational asset than never let them love in the first place? The mixed track record for this question pointed to a potential cure. In instances where a recreational asset was merely taken away (such as a ski hill in Canada), the blowback could surely offset the goodwill that might have been gained during its original dedication. However, if the recreational asset were moved, it could continue to provide a social benefit. On the Mesabi Range, we were accustomed to moving things for mining. Hibbing's tagline before the rebranding effort described earlier in this chapter was "the Town that Moved," and recent publications have suggested that every town on the Mesabi Range (except for Coleraine and Marble) was either partially or completely moved. The record showed that moving a recreational asset often improved both its utility and the social standing of the company that moved it. In my home territory of Upper Michigan, when cross-country ski trails were moved to accommodate mining expansion, the secretary of the 114-year-old ski club described it as "a blessing in disguise" because the trails could be upgraded to the latest standards. Such relocation might cost some money, but that is where return on investment becomes important: Why wouldn't you spend the additional money if your initial investment was a success? This philosophy bore fruit in the goodwill Erie Mining Company engendered when it promoted the value of multiple resource management. Today, when "social license to operate" is the top concern of mining company executives, reintroducing multiple-use principles by implementing intermediate recreational amenities might be a game changer for the mining industry and the region that supports it.[12]

In the debate over the recreational potential of the Mesabi minelands and the legislation that would enable it, a new player was introduced to the Redhead team. Jeff Schoenbauer was a lanky, whiskered, semiretired landscape architect who had served with St. Louis County commissioner Keith Nelson on the Greater Minnesota Regional Parks and Trails Commission, which approved the use of the state's new recreational monies raised with the 2008 passage of the Legacy Amendment. He had been the principal author of the MDNR's *Trail Planning, Design, and Development Guidelines,* which specified the optimal trail widths and conditions for everything from cross-country skiing to all-terrain vehicles to mountain bikes. Schoenbauer was a mountain biker who drove a Mercedes Sprinter van that had been converted into an RV, complete with a rack full of bikes and a halved canoe that could be bolted together to float any nearby river. Importantly, he had been the owner's representative for the MDNR during construction of Cuyuna's first single-track, dedicated mountain bike trails six or seven years earlier. In that role, he had incorporated landscape design principles into trail building and acquired some hard-won knowledge about how to attract the most talented trail builders. His work had left riders wowed. As the legislation enabling Redhead was being drafted, Nelson suggested that the IRRR hire Schoenbauer to perform an independent study of Redhead's viability and the opportunities it presented.

Schoenbauer's assessment of the potential for destination-quality mountain biking at Redhead, published in July 2017, was optimistic. Like Cuyuna, Redhead had the topography and soils that were ideal for mountain biking: "From a design standpoint, there is little doubt that a high-quality mountain biking facility could be built on the site." He wrote that the liability issues, including those specific to the repurposing of minelands, while a legitimate concern, were not necessarily unique. Other facilities had successfully managed such concerns, including Quarry Park and Nature Preserve in Stearns County, where water-filled granite mining pits and old granite debris piles had been repurposed into hiking and biking trails, swimming areas, and observation platforms. But because of the newness of the proposed changes to Minnesota Statute 180, which would invite the public inside the fence line for the first time, the report advised that Redhead be positioned as a pilot project and then carefully observed for its impact. Schoenbauer concluded, "If the liability issues prove solvable (or at least acceptable) to policy makers, and a prototype development like this actually happens and is successful, the prospect of using these mining lands to help foster regional economic development and quality of life enhancements becomes much more tangible."[13]

He did not see an issue with the possibility that Redhead might need to be relocated if mining were to reemerge: "The fact that mining may one day return to the site is actually not as concerning as it would be for a paved trail or other types of recreational infrastructure. In fact, rerouting of trails and 'playing with the design' is commonly done to keep the riding experience interesting and diverse. So, the potential need to relocate trails over time should not be considered a knock out factor in the decision to move forward with this pilot project."

Ultimately, Schoenbauer's report agreed with our own analysis of Redhead's potential—that it could become a genuine recreational destination and, if the liability issues and fencing law changes were managed, open the door for minelands to improve the quality of life for Rangers and help build our economy. While his report echoed themes Plummer and I had been touting for years, his expertise in the field added an important element of gravitas that was helpful to convince those in positions of power, such as St. Louis County commissioner Nelson and the new IRRR commissioner, Mark Phillips, that our idea was sound. In fact, over the years that followed, Schoenbauer's expertise—in combination with our locally focused advocacy and vision—proved a powerful force in advancing the project through the final stages of design and community acceptance.

The report also suggested that, after nine years of development, Redhead was finally ready to move from a volunteer-led effort to a professional one. An updated LVP innovation grant application was submitted in September 2017, and $28,700 was conditionally approved to formalize the design. That meant that I, as a professional engineer and project manager, could actually spend time during working hours to advance the project instead of reserving it for my evenings, weekends, and vacation days. The first stage of work was to refine the conceptual design and the estimated budget for construction of the Phase 1A, 1B, and 2 trails and to develop plans and specifications for construction of the first trail. The project was done by Barr for the city of Chisholm under funding provided by the IRRR's LVP innovation grant.

■ ■ ■

Fieldwork to design the trail was set for November 2017, when all the leaves had fallen and the sight lines would be at their best. To tackle the challenge, Barr assembled a team consisting of a landscape architect, a mapping specialist, a civil engineer, me as project manager, and a cadre of specialists—including Schoenbauer as well as Minnesota professional trail builders Tim Wegner (who had been instrumental in advocating for, securing funding for, and building Cuyuna) and Adam Harju (who had built trails on Minnesota's North Shore). We were accompanied in the field by Plummer and Erik Olson, an accomplished trail builder in his own right who had spent hundreds of volunteer hours for IROC hand-building mountain bike trails at Lookout Mountain north of Virginia. The idea was to capture the absolute best trail lines that the technically challenging Redhead site had to offer while conferring in the field with experts who could bring together perspectives from landscape architecture, engineering, and trail construction. Being Minnesota in mid-November, of course it snowed. Redhead's steep slopes were coated with white as we donned our blaze orange and our winter boots.[14]

I was profoundly impressed by the design concepts that emerged from the field visit. The landscape architects spoke of vistas that would be purposely hidden until the rider emerged from the woods at the point of maximum visual impact. Schoenbauer shared his philosophy that water glimpsed from a high vantage point through the trunks of large trees was somehow deeply comforting and, therefore, beautiful to humanity. And the Redhead site had such views from every angle. No stranger to earthwork (at this point, I had overseen engineering construction projects for twenty-three years), I was astounded by the slopes the builders would construct upon—chaining their mini excavators to trees if necessary—to bench-cut the trail in the exact right place. My job as an engineer with local knowledge was to steer the group to the "must-see" vantage points the site had to offer (known as positive control points) while keeping us away from areas that were either too dangerous or off-limits due to landownership, engineering, or other environmental concerns (negative control points). As the original visionary for Redhead, it was deeply exciting for me to see the expertise of such a distinguished group applied to the idea Plummer and I had been pursuing for so long. By the end of November, we had documented a preliminary design for the trail that included input from the whole field team plus Marty Halverson, a representative from the

Design of the Redhead trail brought together landscape architects, a mapping specialist, a civil engineer, and professional trail builders. *Author's collection*

Chisholm City Council. We felt we had created a vision that would make Redhead shine.

There were some pivotal developments in late 2017 and early 2018. On the far western edge of the Mesabi Range, another mountain bike club (the Grand Rapids and Itasca Mountain Bike Association, aka GRIMBA) and another city (Cohasset) were designing a mountain bike trail near the former Tioga Mine Pit. And on the eastern flank of the Mesabi, the IRRR had commissioned a study for Giants Ridge, suggesting that proceeds of a 2010 recreational use tax levied in the city of Biwabik and surrounding areas be used to construct both cross-country and gravity-focused mountain biking trails that would make use of the ski hill's chairlifts in the summer. The projects, together with Redhead, would develop recreational trails spanning the entire 120-mile Mesabi Range from west to center to east. These complementary, not competitive, assets would reward visitors with multiple opportunities to capitalize on travel to the Mesabi. We saw these three projects as a trifecta and began to unofficially refer to them as the Laurentian mountain biking venues.

Jeff Schoenbauer immediately grasped the benefit of pursuing all three projects simultaneously. With these routes bundled together, what might have been seen by the country's best trail builders as an out-of-the-way project suddenly began to look like a significant, multiyear contract opportunity that was worth the drive. By attracting high-quality talent, the trails could be built to their full potential quickly—three destination systems constituting roughly seventy-five miles of new trails over just two or three years—and allow the Mesabi Range to explode onto the mountain biking scene rather than timidly step in (perhaps too late to capitalize on the sport's rise in popularity). Schoenbauer's mantra was "You don't build a golf course just one hole at a time—you build the whole thing!" In order to meet this short timeline, many builders would have to be working simultaneously. The trail builders were viewed as artists who could leave their signature imprint on the projects and construct the areas of the trail best suited to their individual skill sets, whether that was natural-looking, narrow uphill climbing routes that maximized the beauty of the transitioning mine landscapes or swooping, semicircular downhill bermed sections that rode more like a roller coaster than a bike trail.

The trifecta was an ambitious vision, one that had never been tried before (at least in the Midwest), and it was admittedly grander than what I had ever conceived for Redhead alone. But it made so much sense to try to capitalize on the economies of scale that building three trails at once would provide, as well as their ability to attract the industry's top talent. We decided to pursue the trifecta. Go big or go home. Build it all at once. The full-throated support of IRRR commissioner Mark Phillips, who boldly endorsed our vision, proved key to this proposal.

Our team began to calculate the costs required to build all three destination venues at once. The investments were sizeable based on the concept plans that had been developed for each—$1.777 million for Redhead, $2.548 million for Giants Ridge, and another $0.625 million for Tioga (atop the original $1.3 million that had been secured through other grants)—but justifiable in terms of the return-on-investment framework presented earlier. Together we decided to ask the IRRR board for the entire $4.95 million in one request.

On January 24, 2018, I went before the elected officials who constituted the IRRR board with a deck of eight PowerPoint slides laying out the vision for the trifecta of destination-quality mountain bike parks on the Mesabi Range. I was introduced as both a volunteer officer of IROC and an environmental engineer with Barr. It seemed like a small but crowded room of perhaps forty or so attendees. Speaking for about ten minutes, I explained how the projects were consistent with the Laurentian Vision, how mountain biking was already proven in our region at both Duluth and Cuyuna, and how perfectly mineland suited the needs of natural-surface trail construction. Representatives from the Cuyuna Range echoed the overwhelmingly positive impact mountain biking had had on their region. Only when I received a question from Minnesota senator Tom Bakk that had been texted to him by the *Ely Echo* did I begin to suspect my audience was larger than those in the room. But not until the next day—when a colleague from Barr in Minneapolis greeted me by saying, "I heard your voice on Minnesota Public Radio driving to work this morning"—did it fully dawn on me that my speech had been simulcast and recorded online. To my knowledge, it was the first time interest in Redhead Mountain Bike Park had spread beyond the local population and our small project team.

The IRRR board awarded the full $5 million request to fund the construction of all three destination mountain bike parks. The next day, I received calls from the *Star Tribune*, the *Duluth News Tribune*, and other news outlets interested in knowing more about the IRRR's unprecedented action. But in the atrium of the IRRR office in Eveleth, where most of us who had been involved in the project's development for so long were high-fiving one another, not everyone was happy.

■ ■ ■

By now, Chisholm's mayor, clerk-treasurer/administrator, and city council were different from those individuals in place when the project was enthusiastically supported in 2015. A lot can change politically over the course of three years. And the Redhead project had changed too—from being a ten-mile "local trail" to a twenty-five-plus-mile "destination-quality" trail system. During this period, the local ATV club had contacted the city council to ask that ATVs be allowed to pass *through* the mountain bike park while avoiding the bike trails themselves, an attempt to channel motorized traffic to a legitimate route and thereby, hopefully, minimize potential conflict among trail users. Some city representatives perceived this request as local opposition to the Redhead project (a point of view the ATV club's president denounced) and began to question the city's role and involvement. Within this context, the award of $1.777 million felt like additional pressure over questions not yet fully resolved. In short, some representatives of the city of Chisholm were having second thoughts about Redhead—a situation that would come to a dramatic political resolution.

With financing secured, officials needed to make a final decision. The IRRR would construct the mountain bike park, the Minnesota Discovery Center had agreed to operate the trailhead, and IROC volunteers would help the city inspect and maintain the trail, but the city needed to commit to ultimately own and operate the trail system. The arrangement would be similar to that of the ninety-mile Duluth Traverse, which was insured, owned, and operated by the city of Duluth. However, unlike Duluth, Chisholm had fewer than five thousand residents and qualified as an area of environmental justice concern for its high levels of poverty. With city budgets already painfully thin, would the payback for operating and maintaining a destination-quality mountain bike park—whether through increases in tax base or indirect benefits such as higher use of the city-owned campground—justify the costs? It helped that the original funding package for Redhead reserved approximately 8 percent of its construction budget as warranty funds for the first few years of professional maintenance and repairs. Therefore, the city would not need to bear the costs for postconstruction improvements or rerouting that occurred during the critical first few years when the trail was being worn in. But it would need to accept long-term ownership and operation duties; indeed, this stipulation may have caused the mayor of Redhead's first municipal partner to storm out of the meeting room. And the question loomed: Was there actual popular opposition to the mountain bike park, or was that just an overblown misunderstanding of a legitimately voiced concern over the terms for multiple uses of the park?[15]

At the time, the city of Chisholm was led by an energetic mayor named Mary Jo Rahja. She had spiked gray hair and an ever-present twinkle in her eye and spoke loudly enough that it was prudent to tilt the phone away from one's ear when she called. Rahja had been a community organizer pushing for the development of a park around the Bruce Mine headframe and other improvement projects around town. She loved Redhead and called it her baby. For the project to proceed, the city of Chisholm needed to approve a series of land use agreements and a joint powers agreement with the IRRR specifying the terms for accepting the park once it was completed. Mayor Rahja believed in the quality-of-life improvements and the economic development potential Redhead would create.

The city council's vote was set to take place on April 25, 2018. IROC members packed the audience section of the council chambers to visually demonstrate support for the project. Steve Cook, both an IROC member and a former city councilor, provided a history of the project. Joe Sacco, the Iron Range's longest-serving mountain bicycling advocate, stated during public comments that he had "waited [his] whole life for this day to come." The tension was palpable between the factions that supported the bike park and those that did not. It was rumored that state elected officials had made some phone calls to council members the night before the meeting. When the vote to approve the joint powers agreement and land deals that would allow Redhead to proceed came up, the tally was 5–1. The motion carried in favor of Redhead Mountain Bike Park. A second motion, to support development of an ATV trail either through or passing by the park, was also introduced and approved.[16]

Members of the Iron Range Off-Road Cyclists (IROC) celebrated outside Chisholm City Hall when the final votes needed for Redhead Mountain Bike Park were affirmed by the city council, 2018. *Author's collection*

Mayor Rahja, together with councilor Marty Halverson, had navigated Redhead through some treacherous political waters. It was a tremendous demonstration of commitment by the city, political leadership by its elected officials, and rejection of the fear of change that so many Iron Rangers had felt was holding us back in the 2007 branding study survey.

On May 19, 2018, Minnesota governor Mark Dayton signed the changes to Minnesota Statute 180 into effect. Passage of the legislation represented a watershed moment for the future of mineland repurposing on the Mesabi: not only would Redhead, Cuyuna, and the numerous fishing and swimming and diving areas—including Stubler Beach, Buhl Pit, and Lake Ore-be-gone—be legal, but future government-sanctioned mine pit recreation amenities such as ATV trails and other qualified economic development projects might be possible. Twenty years after Jim Swearingen, Dennis Hendricks, John Koepke, and Chris Carlson started to envision the creation of waterfront parks in the pits of the Mesabi, a legal framework finally existed for these projects to begin.

■ ■ ■

The remainder of 2018 and early 2019 were spent on construction preparations, including hiring a construction manager specializing in mountain bike trails (Kay-Linn of Boulder, Colorado) and vetting a pool of qualified trail builders who would simultaneously construct the Tioga, Redhead, and Giants Ridge parks. Jeff Schoenbauer recommended that the contracts be written so that builders were paid for time spent, not miles of trail built, thereby allowing these artisans to take the hours necessary to make the best trails they could imagine on the red palette of the Mesabi Range. Some of the top builders in the country were hired, including Rock Solid Trail Contracting from Michigan and Pathfinder Trail Building from Minnesota. Local trail builders, such as Erik Olson, were also employed in the construction process.

In late 2018, one final, pivotal element fell into place for Redhead. Trail designs up to that point had avoided an area of 120 acres in the center of the park, known as the Pillsbury Lease because it was owned by the Pillsbury family (whose accomplishments are famous throughout Minnesota, from occupying the governor's mansion to starting the eponymous flour company). Dating back to the first years of the twentieth century, the Pillsburys owned the Dunwoody, Glen, and Monroe Mines—from which they received self-described "hefty royalties" and which now occupied a central position in the pit complex where Redhead was to be embedded. The Pillsbury land had been leased to the IRRR in the 1970s for construction of the Minnesota Discovery Center (then known as Ironworld), but the lease restricted use of the land inside the pit rim. Following passage of the exemptions to the mine pit fencing law (and with an insurance and risk management plan in place for Redhead) representatives of the Pillsbury family were approached about constructing mountain bike trails on their lands. Family members residing in Colorado, familiar with the positive impact recreational developments could have for mining communities— many Colorado resort towns, such as Breckenridge, trace their origins back to mining—approved the request, providing a critical linkage through lands with some of the park's most breathtaking vistas. It was a remarkable act of generosity and support from a "fee holder" of mining lands and one of Minnesota's most storied families.[17]

In spring 2019, an environmental assessment worksheet was written for Redhead. Part of the process involved a survey for species that were endangered, threatened, or of special concern. A tiny, endangered plant was identified: *Botrychium ascendens*, better known as upswept moonwort. The moonworts were given a wide berth by the mountain bike trail so they could remain undisturbed in their preferred post-industrial habitat, at least until natural forest succession occurred.

■ ■ ■

Construction of the Redhead Mountain Bike Park began on Monday, July 22, 2019. After more than forty years of silence, excavators once again turned over Chisholm's red earth (though the diminutive size of these mini excavators would probably have elicited a laugh from the last of the natural iron ore miners). It had been almost exactly a decade since my first auspicious mountain bike ride with Plummer. After the project had suffered two deaths and several harrowing near-death experiences, our teamwork, persistence, and problem-solving had paid off. My journal entry from that day reads, "Redhead started—I cannot believe it! Almost too exhausted to get excited." But by Friday, when Plummer and I inspected the first fractional mile of trail that had been built, we were ecstatic and vigorously shook one another's hands, just as we had on that day in a field near Nashwauk. That night, I wrote, "So, so, so gratifying to see my dream project finally taking shape."

Redhead had a soft opening for the first completed sections of the trail on June 12, 2020. It happened amidst the COVID-19 pandemic, which had the unforeseen impact of driving record numbers of people into outdoor recreation. The timing could not have been more perfect. I was interviewed for television, radio, and print by various local and regional outlets. Most impactful to me was the first foreign-language YouTube video of the Redhead trail, which debuted less than a month after the park's opening—evidence that we had indeed repurposed the Mesabi minelands into a destination-quality attraction that was worth the drive. Later that summer, I was on the trail when I met a rider from suburban Woodbury, Minnesota. In casual conversation, I told him I lived on the Iron Range. "The new mecca of mountain biking," he replied. "You are lucky." It felt like a meaningful retort and a sort of closure to the car salesman's put-down from seventeen years earlier.

On June 26, 2021, all 23.5 miles of trails at Redhead were done and ready for the grand opening. It was attended by news outlets, elected officials, bike and gear retailers, IROC volunteers, food trucks, and hundreds of outdoor enthusiasts, including my eighty-seven-year-old mother, who had insisted on making a special trip from Upper Michigan. Mayor John Champa of Chisholm referred to the project as "a gem of the northland." St. Louis County commissioner Mike Jugovich (the city's former mayor) referred to the "blood, sweat and tears of the mining industry" that had been shed to make the project possible. The president of the Iron Mining Association of Minnesota,

Grand Opening held for Redhead Mountain Bike Park

World class trails are built on reclaimed mine land

BY MARIE TOLONEN
CHISHOLM TRIBUNE PRESS

CHISHOLM — People from across the Iron Range and beyond gathered at the edge of what was once an abandoned mine pit to celebrate a ribbon cutting and grand opening of the Redhead Mountain Bike Park.

The Redhead trail is built on the side of an abandoned mine pit near Minnesota Discovery Center in Chisholm, and is now considered by some cyclists to be a world class mountain

PHOTOS BY MARK SAUER

Riders have some fun while cruising around a sharp turn and hitting an incline while enjoying the Redhead Mountain Bike Park at the Minnesota Discovery Center in Chisholm Saturday. The park was officially opened with a ribbon cutting and ceremony Saturday morning.

Local newspapers celebrated Redhead Mountain Bike Park's grand opening in 2021. The broad representation of public sector, nonprofit, and industry leaders at the event symbolized the community-wide, thirteen-year commitment that brought the project to life. *Mesabi Daily Tribune*

Kelsey Johnson, praised the innovative methods with which Redhead had been conceived and built as an intermediate use of minelands. And IRRR commissioner Mark Phillips rightly stated that it "took a village" to get the project done. The director of St. Louis County's mine inspectors office, Joe Austin, rode his mountain bike through a legal, sign-posted gap in the mine pit fence to experience the exhilaration of the flow trail leading into the pit.

The year it opened, Redhead was the cover photo for an *Outside* magazine story entitled "9 New U.S. Trails You Should Try this Year." The Sierra Club hailed it as an innovation in mineland reclamation for community development. It was featured in a dedicated issue of the local newspaper *Hometown Focus*: "After the Shovels Stop: Mineland Reclamation Brings New Life to the Iron Range." In September 2021, Dan Kraker of Minnesota Public Radio did a story on mountain biking's impact on the Minnesota Iron Ranges, including the creation of twenty new businesses in Crosby and Ironton within the first decade of trail construction at Cuyuna; 180,000 people visit that park every year.[18]

Redhead felt the greatest impact of this influx of interest in the region on October 9 and 10, 2021, when over three thousand riders and their families visited Chisholm for a high school mountain bike race. For those two days, the population of Chisholm nearly doubled. It was rumored that the Chisholm McDonald's ran out of Big Mac buns and the wait times at some locally owned establishments, such as Boom-Town Brewery and Palmer's Tavern in Hibbing, was over two hours. Overnight accommodations were sold out within a fifty-mile radius. Redhead was helping to combat some of the issues Iron Rangers had identified in that community poll so many years earlier: thousands of young people were being attracted to the Iron Range, and these outsiders were being welcomed into the community and helping our retail sector grow and thrive, at least for a few weekends per year.

While tourism impacts can be here one day and gone the next, not unlike the boom-and-bust cycles of mining, the greatest impact of mineland reclamation and repurposing could be attracting and retaining residents. A study conducted in Whitefish, Montana, showed that the tourism appeal of the ski resort was important but did not have nearly the same economic impact as its effect on drawing long-term residents to the area.

In December 2021, I was snowshoeing at Redhead with a group of volunteers who were preparing the trail for winter riding with fat-tire mountain bikes. One of the women in the group, whose husband had taught advanced mathematics to my two oldest children, admitted that they had been on the brink of leaving the Iron Range when Redhead opened. Instead, they had decided to put down a large amount of money to buy one of the Iron Range's iconic structures—the Reed Building, which features three stories of retail and residential space connected via an Art Deco elevator—to create a health center and hostel to welcome visitors to Redhead. They not only bought and preserved an iconic building but opened a series of small businesses that will keep them and others employed in the Iron Range economy. A better story of workforce retention and economic diversification could not be told.[19]

The idea that recreation could diversify and strengthen the struggling Iron Range economy has been pursued for decades, from promotion of fishing resorts to heritage tourism, but until Redhead the focus of these efforts on the Mesabi had rarely been centered on silent-sport adventuring within this minescape. And while it is unlikely that Redhead alone will be the panacea that preserves the economic vitality of the Mesabi Range in the future, early indicators suggest that this pilot project is working. It demonstrates that inactive minelands have value that, when harnessed, can help broaden and sustain the vibrancy of the Iron Range. It also shows that mining need not conflict with recreation and tourism and, in fact, can create unique and interesting landscapes populated by what Cal Flyn calls "nature of the fourth kind"—distinct from pristine, cultured (farmland or managed forest), or ornamental (urban landscapes) spaces and, "in their authenticity and self-direction, . . . a form of wilderness worth preserving in their own right." Redhead was preserving some of this fourth kind of nature, at least temporarily, and already in its first years was attracting young people and positive attention to the unique recovering minescape of the Iron Range. It was strengthening the economy too, and while not on the same scale as billion-dollar mining projects, its positive impacts were happening quickly. The environmental, social, and economic impacts were real. These positive effects were recognized in early 2022 when the Redhead project won three statewide or national-level awards: the Minnesota Brownfields ReScape Innovation Award, the

Environmental Initiative Award for Rural Vitality, and the American Society of Reclamation Sciences film festival award for the short documentary *Reclaimed*. The partnerships and community-building aspects of the project were also celebrated in the documentary *Biketown*, produced and filmed by the national mountain bicycling publication *Freehub Magazine*; the documentary has been viewed by hundreds of thousands of fans and was debuted locally in 2022 to a sellout crowd at the Minnesota Discovery Center.[20]

■ ■ ■

When the Redhead concept was relocated from its original site in Hibbing to its final location in Chisholm, I lost connection with the original LVP charrette drawing, a glimpse at which had offered momentary inspiration early in the project's conception. After the project shifted from where the 2003 charrette drawing had depicted the "Boy Scout Hill Regional Park," I had assumed that Redhead had moved beyond the visions portrayed by the LVP. (The local library no longer holds the LVP charrette drawings in its collection, and so I had no means of knowing what, if anything, had been envisioned for the area south of Chisholm.) Only while researching this book in the quiet, well-lit, and climate-controlled spaces of the reading room of the Elmer L. Andersen Library in Minneapolis did I see on the 2003 charrette drawings that someone had shaded a block of green labeled "Ting Town Mountain Bike Park" just east of where Redhead exists today. In the silence of the reading room, I was floored. How had I not known that a mountain bike park had once been conceived for Chisholm? And which landscape architect had applied the green paint and label "Mountain Bike Park"? What did they envision at the time? And how could I never have met them?

In contemplating this revelation, two thoughts occurred to me. First, I realized that the Redhead Mountain Bike Park had become, next to the Thomas Rukavina Memorial Bridge, the second full-scale manifestation of a Laurentian Vision charrette that had actually been constructed. Two decades after its bumpy beginning, the LVP was beginning to amass a track record I believe would make Jim Swearingen, Dennis Hendricks, John Koepke, Chris Carlson, and all of its original

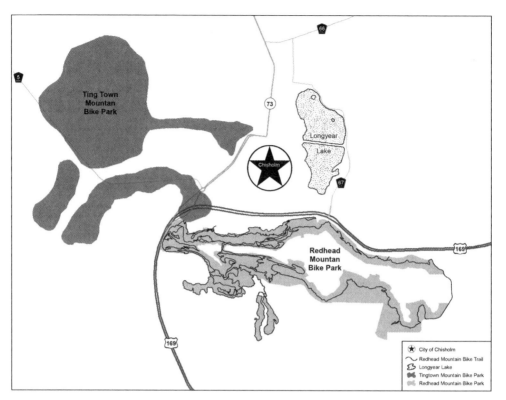

"Ting Town Mountain Bike Park" had been envisioned during the Laurentian Vision charrette in 2003, nearly two decades before Redhead Mountain Bike Park was opened just to the east of the original site. While Ting Town was never developed beyond this drawing, the concept's recurrence demonstrates the interconnectedness of mineland reclamation activities across the generations as well as the decades-long timelines required for projects to come to fruition. *Courtesy of Jim Lind*

visionaries proud. Second, if it took two decades to build the first two LVP projects, it would surely take over a century to complete them all. In this context, the full interconnectedness of the Mesabi Range's mineland reclamation story became clear to me. While this chapter is from my perspective, my part in the Redhead project, which Commissioner Phillips rightly said "took a village," was small in comparison to the overall effort required to succeed. It would take many, many villages and many projects to fulfill the whole mission. Career after career has been spent in the pursuit of mineland reclamation on the Minnesota Iron Ranges, and many more careers will be made in its completion. The task—for all who care about the outcome, whether in the public sector, mining sector, or general populace—as Swearingen said is just getting started and over time will require many hands on the wheel.[21]

Since the amendments to Statute 180, new mountain bike trails have been proposed in Biwabik, Nashwauk, and Mountain Iron. New leaders in the city of Hibbing are enthusiastic, as are several officials for national, state, and local offices who called out Redhead by name during their 2022 campaigns. ATV clubs are building new trails for motorized use that will supplement the "silent sports" trails in a manner similar to Moab's complementary jeeping and mountain biking scenes. Disc golf courses have been built in communities such as Hibbing and Chisholm and include the champion-level course in Buhl. These newer recreational offerings add to the repurposed minelands that predated the change to Minnesota Statute 180, such as the motorized Iron Range Offroad vehicle park and the golf courses at Eagle Ridge and the Quarry at Giants Ridge. The list of recreationally repurposed minelands on the Mesabi is becoming formidable and diverse enough that, collectively, they sound like amenities one might expect of a resort community. A place where, perhaps, one might want to visit, do business, and live—whether connected to the mining industry or not.

Epilogue

This is a history book. While I have labored to explain the thinking of Minnesota's leaders of mineland reclamation through time, their motivations and future-oriented actions are inherently time-stamped in the past. These thoughts and actions belong, rightfully, to the era and circumstances in which they were held and taken, even if in some cases such actions resulted in unintended consequences that those who came later needed to deal with. Every story features someone who was doing their best, with the resources at their disposal, to create state-of-the-art mineland reclamation and repurposing for their time. The stories' trajectories tend to arc upward—whether industrial-scale disturbance to resource conservation (chapter 2); experimental regrowth and reuse (chapter 3); mandatory stabilization and revegetation and local-scale repurposing (chapter 4); long-term regional visioning, public engagement, education, and planning (chapter 5); or temporary but immediate revitalization and its accompanying social, legal, and community developments (chapter 6)—toward a zenith in which the iron lands of the Mesabi exhibit a full second life after mining: perhaps not fully restored, but revived.

However, the story of mineland reclamation and repurposing is far from over. Some, like Jim Swearingen, would say it remains in its infancy. So, given the arc of history, what, if anything, can be said about the future of mineland reclamation and repurposing on Minnesota's iron lands?

The reclamation story will play out upon one of the most dynamic landscapes on Earth, where mountains are literally moved every day, roads that once took you one place suddenly take you somewhere else (or end altogether), and whole towns have been moved or abandoned. While this is happening, the forces of nature will act in ways both violent (mass erosion of unstable, pre-1980 pit walls) and silent (regrowth of volunteer forests or groundwater reentering idled mine pits). This dynamism and interplay between human actions and the forces of nature are inherent in Iron Range life, making it difficult to reliably predict the immediate and medium-term future for the Mesabi. But rather than chafing at the transitional nature of mined lands and their innate impermanence, it may be beneficial to focus attention on the long-term future and the types of intermediate actions that can make sense regardless of the uncertainties involved in a shorter time frame.

In the field of economics, Stein's Law (named after 1970s economist Herbert Stein) states, "If something cannot go on forever, it will stop." While this statement seems self-evident, it applies to the supplies of iron resources on the Mesabi, which are finite. Iron mining will end some-day. The record of historical predictions of that day range from clearly shortsighted (Andrew Carnegie's "twenty-five or forty years," which would have ended Minnesota mining no later than 1948) to highly op-timistic (E. W. Davis's five to six hundred years, which would project taconite mining out to the year 2557). But factors that influence the socioeconomics of mining—such as global supply and demand, techno-logical development, and the social and legal licenses to operate—will likely have an even greater effect on the long-term future of Minnesota iron mining than its finite supply. With all of these unknowns, some of the best options are cautious, resource-based predictions that assume the continuance of stable socioeconomic factors, such as the estimate made by Peter Clevenstine, Minnesota Department of Natural Re-sources division of Lands and Minerals assistant director, in the *Mesabi Daily News* on October 30, 2019: "There is existing taconite, like they are mining today, that would sustain mining for another 100 years."[1]

In other words, taconite mining went from heartbreaker to life giver on the Mesabi Range in 1957, and, while it will someday end, as a regional enterprise it still has a lot of potential life left to give. That means Mesabi iron will be made into one of the cars you may one day buy (maybe an electric car, which requires more steel than a gasoline-powered car) and the high-efficiency, water-saving dishwasher that will clean my reusable water bottle. Along the way, that iron will con-tinue to pay into the Department of Iron Range Resources and Reha-bilitation's tax coffers (helping fund more mineland reclamation and economic diversification for the Iron Range), the University of Minne-sota education that residents' children and grandchildren may receive, the profit margins of the mineral rights holders and mining companies, and the paychecks of Iron Range families like my own.

That being said, the start of mineland reclamation and repurposing is not synchronized with the ending of the enterprise of iron mining as a whole. Contrary to what some people may think, reclamation does not begin when the last taconite pellet or hot briquetted iron nugget is shipped out of Minnesota. It begins on the day when any given shovelful of mined land is disturbed for the last time, which is an inherently local event. For example, the reclamation clock of the Kinney stockpile described in chapter 3 started when the train dumped its last load of overburden in 1902, which allowed that pile to achieve the advanced state of forest succession I witnessed 120 years later. Simultaneously, just a few miles away at Jim Swearingen's and Dennis Hendricks's Minntac, the active mining of taconite was still going strong. The industry term for this side-by-side advancement of mining and reclamation is *progressive reclamation,* and it is the preferred—and even rule-mandated—state of operations. Modern mining and reclamation happen simultaneously, with reclamation work following rapidly on the heels of the last day that any plot of land is touched by human hands.

In the future, the most advanced mining operations in Minnesota will feature the *progressive repurposing* of minelands. For example, over the next sixty-five years at Northshore Mining Company, as certain parts of the pit and stockpiles are completed, they will be systematically constructed into soil and tree nurseries, terrestrial and aqueous habitats, trails, and visitors centers in accordance with the award-winning framework plan developed by John Koepke and Chris Carlson, cofunded by the mining company and IRRR, and ultimately approved by the MDNR. This transformation will occur while other parts of the pit are still being mined and epitomizes the modern concept of a "circular economy," where "little is wasted and reuse reigns." As a second example, the newly developed *intermediate use* concept and legal framework pioneered for the Redhead Mountain Bike Park will allow Iron Range communities to work with willing landowners to develop temporary, government-sanctioned recreational uses of idled minelands, putting them back into productive use rather than simply fencing them off while awaiting the return of economic conditions that make them favorable to mine. If undertaken in the planful, purposeful way envisioned by the Laurentian Vision Partnership, these types of projects could systematically transform the Mesabi minelands into one giant landscape asset in a way that is compatible, and potentially even symbiotic, with the continuation of taconite mining. Taking advantage of the "free energy materials" available during active mining (by earth-moving and tree replanting), some of these projects might be possible at minimal extra cost to the mining companies and landowners while providing them a huge opportunity for community support—the critical social license to operate that mining company CEOs rate as the biggest concern for the future of their industry.[2]

While it remains a $1.2 billion per year enterprise and a corner-stone of the Iron Range economy that is likely to continue for decades, iron mining in Minnesota, through no fault of its own, has undergone a slow decline. Employment and production by the iron mining industry peaked decades ago. And the number of Minnesota's taconite-producing plants has fallen from a maximum of eight down to six, meaning it is unlikely that we will ever again achieve those historic highs in output because the infrastructure to do so has been removed and would cost billions of dollars to replace. All of the taconite facilities producing today opened in the 1970s or earlier. And at the time of this writing, no scram iron mining operations have produced iron ore concentrate since Magnetation went bankrupt in 2015. Essar Steel's combined iron mining and steelmaking operation (whose ground-breaking was described in chapter 6) would have been the first major taconite operation built since the Mineland Reclamation Rules were passed in 1980; however, because of changes in economic conditions during construction, it lies half built and mired in debt just north of Nashwauk. Thus, while iron mining will continue—likely for decades or centuries into the future—the historical trend indicates that few to no new plants are likely to be built and previous levels of production and economic impact are unlikely to return. This assessment, in my opinion, means two things for the future of the Iron Range: 1) Minnesota's existing taconite plants should be supported and protected as they jockey for position in tomorrow's iron ore marketplace, with its attendant socioeconomic conditions, and 2) we should not close our eyes to the inevitable ending of Minnesota iron mining and should redouble our efforts to use its ongoing economic power to create a vibrant and diversified future for the Iron Range.

That is where the potential of future mineland repurposing comes in. Planful mineland repurposing can be symbiotic with active mining by growing the recreation and tourism industries created on repurposed minelands, as shown at the Redhead, Tioga, and Cuyuna mountain bike parks; the Gilbert off-highway vehicle park; the Eagle Ridge and Giants Ridge Quarry golf courses; the scuba-diving attractions of Lake Ore-be-gone; and other similar examples of revitalized minelands. Such amenities can be brought online much quicker, and with less controversy and uncertainty, than new heavy industrial projects. This point was made by city of Cohasset administrator Max Peters, who was instrumental in creating the Tioga mountain bike trail system and, on the day I spoke with him, was served in a lawsuit related to an environmental assessment completed by the city for a new large industrial project coming to town. Peters, whose motivation is simply to ensure a solid tax base and good jobs for his community into the future, said, "If new industry is going to be this difficult, I will turn toward recreation and tourism. I wish I could just build Tioga again. I would build it again ten out of ten times."

A recent study in *Mining Engineering*, "Accepting the Depletability of Deposits and the Creation of Ghost Towns," primarily discusses civilization's need for minerals, the finite nature of mineral deposits, and possible solutions to mitigate the social and environmental impacts of mining. Particularly intriguing is a table documenting the modern status of eighteen small, remote mining camps in Colorado where mining had been discontinued. The present incarnations of each of these towns fell into four categories: ski towns (including Aspen, Telluride, and Breckenridge); tourist towns (such as Lake City and Silverton); gambling towns; and ghost towns. If these same categories were transferred to rural northeastern Minnesota, where gambling is illegal and recreation and tourism are inextricably linked, they would leave just two opposite outcomes for towns after mining: recreation/tourist town or ghost town. But the future of Mesabi Range towns need not be so stark—after all, we have now diversified our economy to include significant health care and financial industries, the nation's second-largest solar panel plant (Heliene), precision parts and paint jobs for aircraft (Hibbing Fabricators and Midwest Aircraft Refinishing), and industrial-sized radiators (L&M Radiator). However, the point is sobering. After mining, we must change or die. That is the future for a place whose own occupants rate fear of change as one of their top three challenges.[3]

In 1916, Edmond DeLestry denigrated the Iron Range as a place where "all the natural ornamentation disappeared long ago and only huge mounds of debris surround the community wherever the eye may rest. . . . [W]hatever of attractiveness they would ever own, must perforce be artificially created." It is true that the Mesabi, unlike the former mining districts in Colorado's ski towns or Michigan's Marquette Iron Range, does not possess a naturally beautiful snow-capped mountain range or a Great Lake. In the earliest booming days, attempts to create artificial attractions to make the Mesabi livable led to architectural wonders such as the city halls, libraries, public schools, fountains, and city parks that still enrich our communities. But like new iron processing plants, almost no new civic wonders are being built, and those we have are sometimes lost because they are costly and difficult to maintain.[4]

This is the second way mineland reclamation and repurposing can have a positive impact on the future of the Iron Range. In the century since DeLestry wrote and in the centuries to come, the landscape has been and will be transforming from his "mounds of debris" into a place with a beauty of its own—Cal Flyn's *jolie laide*—where natural and anthropogenic activities blend to create a unique and unexpected form of wilderness. Unlike the civic architecture of the Iron Range, which requires constant upkeep, nature can create, maintain, and increase this form of attractiveness, if we humans give its workspace proper forethought and nature itself a decent start. The resulting ramshackle wilderness will emerge slowly over the decades that reclamation takes, temporarily giving shelter to rare and threatened pioneer species. That unique form of wilderness can catch people off guard, leading them to take notice and come back to watch the future unfold. Bob Dylan wrote of the Mesabi, "What was in the air here is still here. [It's] still untarnished." At that time, the minescape already existed, so he couldn't have been speaking of an unadulterated Iron Range. He was speaking of the magic of the Mesabi that somehow fosters creativity and an ability to renew itself.

Fifteen years ago, as Minnesotans and Iron Rangers were contemplating a potential uptick in mining this part of the state (one that has not fully materialized, as described in chapter 6), the Minnesota Center for Environmental Advocacy (MCEA) and the Sierra Club funded a report entitled *The Economic Role of Metal Mining in Minnesota: Past, Present and Future* by Thomas Michael Power from the University of

Montana. The report made several points, including that "people and businesses have pursued what they perceive to be higher quality living environments"; that amenities (such as "an attractive landscape and climatic features") can bring in "recreationists, retirees and other new residents"; that "extractive industry, including metal mining, *by itself,* has generated ghost towns in the past" (emphasis added); and that "high quality living environments, on the other hand, prevent ghost towns by holding and attracting economic activity." Power states that "damage to the landscape and waters of northeastern Minnesota *could* close off this source of economic vitality and leave the region entirely dependent on volatile international metal markets" and that metal mining is a "landscape-intensive activity *that almost always has had* significant negative impacts on the natural environment" (emphasis added).[5]

At face value, these statements are not incorrect. This book starts with the industrial-scale disturbance that was wrought upon the lands of the Mesabi before there were laws and social pressures to reclaim minelands. But Power's report was written before places such as our mountain bike trails, ATV parks, and other mineland repurposing projects had demonstrated that minelands could be used *to create* landscape amenities that make for a high-quality living environment that attracts recreationists, retirees, and other new residents, just as the proponents of the Laurentian Vision advocated. And while it is true that mining *by itself* is landscape intensive and *could* limit other sources of economic activity deriving from a high-quality natural environment, it also, when paired with planful, creative, and environmentally sensitive mine reclamation and repurposing, can be symbiotic and work together on the landscape to avoid such a bleak economic and environmental future for mining communities. As is shown throughout this book, understanding the potential for the Iron Range to be beautifully reclaimed and repurposed to provide for quality-of-life improvements and a diversified economy is not new. Sam Dickinson, Dave Youngman, Jim Swearingen, Dennis Hendricks, John Koepke, Chris Carlson, and countless MDNR, IRRR, and mining company officials have been working on this concept for at least fifty years. With the recent popularity of some high-profile, destination-quality mineland repurposing projects, the public became aware of this potential as well. In fact, recent reporting by the Sierra Club and *Outside* magazine (both of which mention Redhead Mountain Bike Park by name) has highlighted the community-enhancing potential of mineland repurposing and the importance of developing recreation on postindustrial lands as a way to avoid environmental damage to natural wilderness areas, such as the Boundary Waters Canoe Area Wilderness—one effect of a COVID-related upsurge in outdoor recreation. When the BWCAW needs to restrict permit access to protect against environmental damage from high usership, as was the case in 2022, temporarily repurposed recreational minelands will stand ready to meet the demand.[6]

The newly reinvigorated potential of mineland repurposing, at least for recreational uses symbiotic with future mining, is best characterized by a recent conversation I had with Rod Hunt. Hunt is a Korean War veteran, a seven-time published author, the CEO of two current mining companies (one in Canada and one in Alaska), and an original cofounder of Magnetation. Having personally benefited from a lifetime in mining, he has recently created a nonprofit organization called Wolf Head Recovery and Discovery whose mission is, in his words, to turn the Mesabi's "wastelands into parklands." Hunt's vision is to create a kind of interactive parkland that mixes the unnatural, reclaimed beauty of the Mesabi with adventure by creating additional naturally surfaced trails, zip lines, waterfalls, and carbon-storing forests of deciduous trees—kind of like a northern version of Xel-Há, Mexico's commercial aquatic theme park and ecotourism destination. It is to be a parkland dedicated to environmental restoration at the gateway to northeast Minnesota's existing wilderness landscapes, the BWCAW and Voyageurs National Park. When I asked Hunt why, at the age of ninety-seven, he has embarked on such a project, he said, "All I have left is the desire to do something decent." In other words, he wants to give back to the people and the lands of the Mesabi. With tens of thousands of visitors already coming to the reclaimed Mesabi minelands for recreation, he hopes to systematically turn *all* of the postmining Mesabi into parklands over the course of several decades. He hopes that his nonprofit will establish partnerships with mining companies and landowners who, like him, feel a moral imperative to create something "decent" on the Mesabi after these lands have given so much iron and so much wealth.

In that 2007 report for the MCEA and Sierra Club, Power also stated, "Returning the land to its original condition is almost impossible, and even effectively reclaiming it so that natural processes can heal the disturbances can be difficult, costly and ultimately unsuccessful." Once again, at face value, these statements are not untrue. The Mesabi lands will never again look like they did before mining. However, one thing I learned while writing this book is that our assumptions of "original condition" can be inflated or easily misunderstood. For example, the "untouched" wilderness that European settlers saw in the Americas had in fact been occupied and cultivated by Indigenous peoples for millennia (populations that would be decimated by disease after Columbian contact). Similarly, the woods around Henry David Thoreau's Walden Pond were not virgin forest but farmland transitioning back to forestland after having been the nation's breadbasket prior to large-scale agriculture developing in the Midwest. The point is that nature is ever-changing such that almost nowhere is it in its original condition. While the Mesabi will never be the same as it was before mining, in the future more of it could look like the Kinney stockpile—not original, but wild, and functioning in a way that positively contributes to the region's health.[7]

Power is correct that reclamation and repurposing of abandoned minelands are slow, difficult undertakings and that, as is demonstrated throughout this book, reclamation experiments sometimes fail. As noted in chapter 5, future actions by the mines operating today—even if they follow the Laurentian Vision principles to maximize the use of "free energy materials"—will require decades to build as those companies mine their reserves and slowly transition lands into a second use. And while the rules require a minimum standard of reclamation for today's taconite mines, they do not mandate or provide funding to reclaim the ninety years' worth of mined lands created before the rules came into effect, nor do they require full ecological restoration or repurposing of mined lands. Just implementing the Thomas Rukavina Memorial Bridge and the Redhead Mountain Bike Park cost $230 million and $1.777 million, respectively—and for the bridge, $15 million of the cost was paid to the mineral rights holders to ensure that the property beneath the bridge abutments and roadway would not be mined in the future. To date, these high costs for reclamation and repurposing, beyond what is required by the rules, have made it difficult to get started on an Iron Range–wide transformation such as that envisioned by Rod Hunt.[8]

But the ability and urgency to embrace mineland reclamation and repurposing will only continue to grow. Redhead Mountain Bike Park and similar developments demonstrate that interim repurposing of minelands can be done at relatively low cost, need not encumber the potential for future mining in the region, and can be executed within the legal framework of government-sanctioned recreational use. Under commissioner Mark Phillips, the IRRR recently established a regional trails program that puts millions of dollars per year into the development of naturally surfaced trails throughout its service area. At the federal level, the 2022 infrastructure act recognizes that significant investments are needed for reclamation and repurposing of abandoned minelands, providing $11.3 billion over a fifteen-year period (although most of this funding will be directed to reclamation of coal mines, not metal mines). And sustainability and social license to operate initiatives have led to private investments into similar kinds of regional initiatives, such as the Walton Family Foundation's $74 million stake in making Bentonville, Arkansas, one of the world's top mountain biking centers.

When another taconite plant closes, the diversity of the Iron Range's economic base—or lack thereof—will (again) become a community issue. History suggests such events may occur only every two or three decades, just slowly enough to slip from our minds. But short-term and medium-term choices are important. Iron Rangers have become pretty comfortable living on the minescape. While some love it and others hate it, the ubiquitous presence of the mine dumps on everything from politicians' advertisements to Minnesota North College's new logo demonstrates that it is a permanent part of the Iron Range's physical and cultural landscape. The choices we make in this landscape will create the future for this region.

Even if Rod Hunt lives to be the oldest person on Earth, he is unlikely to see the whole Mesabi Range wasteland turn into a parkland. For now, however, he is on top of the world and pursuing his passion for this project so intensely that his last email to me ended with the sign-off "My pants are on fire!" This level of enthusiasm could sweep through Iron Range communities. This book shows that the underpinnings for such endeavors are available in the historical record—from technical studies on reforestation inspiring regional landscape plans to legal mechanisms for intermediate recreational repurposing that preserves current and future mining opportunities. To be clear, the Mesabi Range, at 140,000 acres, would be the largest contiguous, coordinated mineland reclamation and repurposing project on Earth. It would require innovation from the land that has already given rise to the genius of John C. Greenway, Sam Dickinson, and Bob Dylan. It would require funding from federal, state, and private sources to the tune of tens to hundreds of millions of dollars. It would require unprecedented buy-in at all levels—landowners, mining companies, nonprofit organizations, elected officials, public servants—which is not entirely dissimilar to what is happening just seventy-five miles downstream in the active, industrial ports of Duluth and Superior, where the St. Louis River area of concern cleanup and restoration is being jointly led by the cities and the US Army Corps of Engineers, the Environmental Protection Agency, the Fish and Wildlife Service, the National Oceanic and Atmospheric Administration, the Minnesota and Wisconsin Departments of Natural Resources, the Minnesota Pollution Control Agency, the Fond du Lac Band of Lake Superior Chippewa, and the nonprofit St. Louis River Alliance. Above all, it would require the kind of political leadership and cross-party coordination that might return the Iron Range to the level of prestige that inspires presidential visits. Given the past achievements that have occurred here, it's not impossible. These undertakings would employ a lot of people on the Range. Ultimately, a Range-wide reclamation and repurposing project may even transition people's mindsets from one of fall, where our most productive days are behind us, to one of spring, where something new is growing on a once cold land. The pursuit may give us and our lands a second life. After all, it was an Iron Ranger—and Nobel Prize winner—who understood, and told the world, that "he not busy being born is busy dying."

Acknowledgments

This book would have been impossible without the priceless contributions of so, so many dedicated people and institutions. In no particular order, they include:

- My editor and de facto writing coach, Shannon Pennefeather of the Minnesota Historical Society Press, her helpful intern Em Poupart, and wonderful freelance copyeditor Madeleine Vasaly.
- All whom I interviewed or quoted from face-to-face interactions, including Mike MacFarlane, José Castillo, John Dougherty, Dave Youngman, Dan Jordan, Julie Jordan, Phil Solseng, Jim Swearingen, Dennis Hendricks, Christine Carlson, John Koepke, Kathryn Ryan, Irina Fursman, Jim Plummer, Max Peters, and Rod Hunt.
- All who reviewed my material and/or checked my facts, including Sandy Karnowski, Chrissy Bartovich, Stephanie Kraynick, Lisa Klaphake, Jim Plummer, Linda Johnson, John Borovsky, Rod Hunt, Max Peters, Christine Carlson, John Koepke, John Baeten, Aaron Brown, Stephanie Skraba, Scott Sands, and Tani Hemmila.
- All who created or collected the figures and images in the book, including Vance Gellert, Christopher Welter, Lilah Crowe, Erica Zubich, Jim Lind, Louise Lundin, Kent Dickinson, Jim Scott, Cheryll Fong, John Koepke, Ardy Nurmi-Wilberg, Joe Treleven, Jon Grinney, Hibbing Chamber of Commerce, and Barr Engineering Co.
- The institutions whose professional contributions made this work possible: Barr Engineering Co., Minnesota Discovery Center, Itasca County Historical Society, Hibbing Historical Society, St. Louis County Historical Society, Ajo Historical Society, Arizona Historical Society, University of Minnesota's Northwest Architectural Archives, Minnesota Department of Iron Range Resources and Rehabilitation, *Mesabi Tribune, Freehub Magazine,* Iron Range Off-Road Cyclists, and the Minnesota History Center.

- All of the practitioners of mineland reclamation and repurposing in Minnesota whose work made this book possible, including those whom I did not specifically identify by name.

And, most importantly, my family, friends, and coworkers who both encouraged me and stuck with me during the four-year journey to bring this book to life, especially Miriam.

Appendix 1

Nonminer Deaths and Injuries in Open Mine Pits Related to Trespassing through Mine Pit Fencing

CIRCUMSTANCE SURROUNDING ACCIDENT	AGE	DATE OF ACCIDENT	NATURE OF INJURY OR CAUSE OF DEATH	TYPE OF ACCESS USED TO ENTER PIT	LOCATION OF ACCIDENT	RESULTING LIABILITY CLAIMS
Group of boys on a raft—Suburban	12 (male)	unknown	Drown	Through fence	Gilbert Mine Pit	Unknown
Path through fence hole to berry patch—Townsite	6 (male)	1973	Drown	On foot through fence hole	Spruce Mine Pit—Eveleth	Yes, but unknown
Playing at mine pit—Townsite, two brothers	7 and 9 (males)	1958	Drown	On foot through fence	Wisstar Mine Pit—McKinley	Unknown
Swimming and scuba diving at pit access road—Suburban	21 (male)	1983	Drown	Fence removed from access road	Embarrass Mine Pit—Aurora	Unknown
Swimming and fishing—Suburban	19 (male)	1985	Drown	Fence removed from access road	Embarrass Mine Pit—Aurora	Unknown
Possible seizure; drove auto through fence into pit—Townsite	41 (male)	1985	Drown	Drove auto through fence	Miller-Mohawk Mine Pit—Aurora	Unknown
Fell down embankment into water—Townsite	6 (male)	1965	Drown	Through hole in fence	Rouchleau Mine Pit—Virginia	Unknown
Went through fence on foot in winter—Townsite	30 (male)	1985	Death due to exposure	Went through fence	Rouchleau Mine Pit—Virginia	Unknown
Fell down bank—Townsite	7 (male)	1976	Broken wrist and bruises	Went through hole in fence	Rouchleau Mine Pit—Virginia	Company 70%; Boy 30%
Walked through fence and over embankment—Townsite	25 (female)	1975	Bruises and fractures	Went through fence and fell down pit bank	Rouchleau Mine Pit—Virginia	Unknown
Fence out for three-wheeler; mesh gone; barbed wire remained—Townsite	45 (male)	1985	Cuts on face, scalp	Jogging and went through fence hole, hit barbed wire at head level	Rouchleau Mine Pit—Virginia	Yes, but unknown

Source: Data retained by the Iron Range Resources and Rehabilitation Board regarding accidental deaths and injuries that occurred from people trespassing through mine fencing, 1958 to 1985

Appendix 2

Early Participants in the Laurentian Vision Partnership (LVP) Working Group

Kevin Adolfs, Minnesota Department of Transportation

Nancy Aronson Norr, Minnesota Power

Jason Aune, University of Minnesota, Department of Landscape Architecture

Tony Bauer, Bauer-Ford Reclamation

Bob Berglund, Northshore Mining Company

Wayne Brandt, Minnesota Forest Industries

Bob Bratulich, United Steelworkers of America

Bill Brice, Minnesota Department of Natural Resources, Division of Land and Minerals

Christine Carlson, University of Minnesota, Department of Landscape Architecture

Dave Carlstrom, City of Chisholm

Gary Cerkvenik, National Steel Pellet Company

Josh Cerra, University of Minnesota, Department of Landscape Architecture

David Chmielewski, LHB Engineers and Architects

Steve Dewar, Minnesota Department of Natural Resources, Division of Lands and Minerals

Samuel Dickinson, Western Mesabi Mine Planning Board

Jean Dolensek, Iron Range Resources and Rehabilitation Board

Jerry Dombeck, U.S. Steel

Jerry Drong, Drong & Associates

Steve Durrant, URS Corporation

David Epperly, St. Louis County Land Development

Bill Everett, Hibbing Taconite Company

John Fedo, Fedo & Associates

Lory Fedo, Hibbing Chamber of Commerce

Carlos Fernandez, University of Minnesota, Department of Landscape Architecture

Donald Fosnacht, University of Minnesota, Natural Resources Research Institute

David Foster, United Steelworkers of America

Latisha Gietzen, National Steel Pellet Company

Ann Glumac, Iron Mining Association

Todd Halunen, URS Corporation

Jeff Hammerlind, EVTAC Mining Company

Mirja Hanson, facilitator

Dennis Hendricks, USX—Minnesota Ore Operations (Minntac), Northern Lands and Minerals

Doug Hildenbrand, Architectural Resources, Inc.

Howard Hilshorst, EVTAC Mining Company

Brian Hiti, Iron Range Resources and Rehabilitation Board

Jonathan Holmes, Ispat Inland Mining Company

Tom Houghtaling, Minnesota Power

Rod Ikola

Anne Jacodisits, Minnesota Department of Natural Resources, Division of Lands and Minerals

Dan Jordan, Iron Range Resources and Rehabilitation Board, Mineland Reclamation Division

Julie Jordan, Minnesota Department of Natural Resources, Division of Lands and Minerals

Bruce Kniivila, USX—Minnesota Ore
Operations (Minntac)

Arlo Knoll, Minnesota Department of
Natural Resources, Division of Lands
and Minerals

John Koepke, University of Minnesota,
Department of Landscape Architecture

Dennis Koschak, LTV Steel Mining
Company

Mike Lalich, University of Minnesota,
Natural Resources Research Institute

Jack LaVoy, Iron Range Resources and
Rehabilitation Board

Doug Learmont

Dave Lotti, Western Mesabi Mine Planning
Board

Jack Lund, City of Hibbing

Dave Maki, Balkan Township

Rick Maki, EVTAC Mining Company

Katherine Martin, University of Minnesota
Duluth

Roger Martin, University of Minnesota,
Department of Landscape Architecture

David Meineke, Meriden Engineering LLC

Steve Mekkes, ArcelorMittal Minorca Mine

Steve Moddemeyer

Eric Mustonen

Duane Northagen, City of Hibbing

Paul Olson, Blandin Foundation

Frank Ongaro, Range Association of
Municipalities and Schools (later with
the Iron Mining Association)

Dennis Oost

Rod Otterness, City of Buhl

Peter Pastika, Northshore Mining
Company

Tim Pastika, Minnesota Department of
Natural Resources, Division of Lands
and Minerals

Tony Pekovitch, Minnesota Power

Tom Peluso, National Steel Pellet
Company

James Pettinari, University of Oregon,
Urban Architecture Program

David Pitt, University of Minnesota,
Department of Landscape Architecture

Paul Pojar, Minnesota Department of
Natural Resources, Division of Lands
and Minerals

Tom Pustovar, Minnesota Power

Tom Renier, Northland Foundation

Kathryn Ryan, URS Corporation

Shelly Sallee, Iron Range Resources and
Rehabilitation Board

Joe Samargia, Summit Consulting

Larry Schmelzer, National Steel Pellet
Company

Herb Sellers, Great Scott Township

Joe Sertich, Northeast Higher Education
District

Jerry Shapins

Dave Skolasinki, Cliffs Mining Services
Company

Vicki Spragg, Arrowhead Regional
Development Commission

Ray Svatos, Iron Range Resources and
Rehabilitation Board, Mineland
Reclamation Division

Jim Swearingen, USX—Minnesota Ore
Operations (Minntac)

John Swift, Iron Range Resources and
Rehabilitation Board

Mike Thomas

Leo Trunt, Western Mesabi Mine Planning
Board

Jack Tuomi, Hibbing Taconite Company

Marty Vadis, Minnesota Department of
Natural Resources, Division of Lands
and Minerals

Pete VanDelinder, Hibbing Taconite
Company

Augustine Wong, URS Corporation

Fred Young

Source: Based on U.S. Steel files and LVP documents stored at the Northwest Architectural Archives,
Elmer L. Andersen Library, University of Minnesota, Minneapolis

Notes

Note to Land Acknowledgment

1. "Biiwaabik," Ojibwe People's Dictionary, https://ojibwe.lib.umn. edu/main-entry/biiwaabik-ni; "Missabe," Ojibwe People's Dictionary, https://ojibwe.lib.umn.edu/main-entry/misaabe-na.

Notes to Chapter 1: Introduction

1. It is unclear whether this 1936 photo depicting the "dumping of waste rock from thirty yard air-dump cars" from Hibbing's Hull-Rust pit represents Pill Hill or one of several other dump locations. Nonetheless, the rail dumping technique is representative of the manner in which the Winston and Dear Dump would have been built.
2. Aaron Brown, "Mr. Power's Undaunted Fighting Spirit," Minnesota Brown: Modern Life in Northern Minnesota, June 8, 2015, http://minnesotabrown.com/2015/06/mr-powers-undaunted-fighting-spirit.html; Edmond L. DeLestry, "The Romance of a Town: The Story of Hibbing, Wonder-Village of the World," *Western Magazine* 7, no. 6 (May 1916).
3. John Dougherty, "From Hibbing Ski Hill, to the World's," Years of Yore, *Hibbing Daily Tribune*, November 18, 2018. Photo is from 1938: mndigital.org.
4. "Program for Community Improvement, Hibbing, Minnesota," 1960, Hibbing City Collection, Minnesota Discovery Center, Chisholm, MN.
5. Barbara Nelson, "Mining Land Use: Level I Report," May 9, 1978, Minnesota Legislative Reference Library, http://www.leg.state.mn.us/lrl/lrl.asp.
6. Edmund J. Longyear, *Mesabi Pioneer: Reminiscences of Edmund J. Longyear*, ed. Grace Lee Nute (St. Paul: Minnesota Historical Society, 1951), 30.
7. At the time of publication, the nonferrous mining permits were undergoing legal challenge.
8. Handwritten quotation on the cover of the 1936 publication *Wonderland of Lake Superior*, Iron Range Research Center, Minnesota Discovery

Center, Chisholm, MN. The stories contained in this book are factual, in that they are supported by research and citations, but are not intended to represent a complete history. In the latter chapters, in which some elements of my personal involvement appear, the events are told from my perspective and are true as I remember them, but others may have differing perspectives or recollections.

Notes to Chapter 2: The Boom

1. Edmund Morris, *The Rise of Theodore Roosevelt* (New York: Random House, 2010).
2. John Campbell Greenway to Theodore Roosevelt, January 22, 1902, Theodore Roosevelt Center, Dickinson State University, Dickinson, ND.
3. Douglas Brinkley, *The Wilderness Warrior: Theodore Roosevelt and the Crusade for America* (New York: Harper, 2009).
4. J. S. Pardee, "Of the Womb of Wealth," manuscript, p. 1, box 3, Oliver Iron Mining Company research files, Minnesota Historical Society, St. Paul.
5. The simplicity regarding "direct ship" ore applies only to mining technique itself and omits the massive undertakings required to develop the infrastructure in northeast Minnesota to excavate and ship iron ore to market with blast furnaces.
6. John McPhee, *Annals of the Former World* (New York: Farrar, Straus and Giroux, 1998), 634.
7. Pardee, "Of the Womb of Wealth," 4. Greenway, during the course of mining the western Mesabi, amassed a personal fossil collection that is now housed at the high school in Minnesota that bears his name. Other fossils found in the area include a crocodile jaw, sawfish teeth, snails, oysters, and petrified pine cones.
8. Pardee, "Of the Womb of Wealth," 5.
9. *Duluth Evening Herald,* February 27, 1892, 3, cited in Fremont P. Wirth, *The Discovery and Exploitation of the Minnesota Iron Lands* (Cedar Rapids, IA: Torch Press, 1937); Andrew Carnegie, speech, Governors' Conference on the Conservation of Natural Resources, White House, 1908.
10. Rukard Hurd, *Iron Ore Manual of the Lake Superior District, with Values Based on 1911 Prices . . .* (St. Paul, MN: F. M. Catlin, 1911), 39.
11. Jeffrey T. Manuel, *Taconite Dreams: The Struggle to Sustain Mining on Minnesota's Iron Range, 1915–2000* (Minneapolis: University of Minnesota Press, 2015), 175.
12. Roosevelt would appoint Greenway to the Panama Canal commission on November 27, 1907. Robert Marsh Jr., *Steam Shovel Mining, Including*

a Consideration of Electric Shovels and Other Power Excavators in Open-Pit Mining (New York: McGraw-Hill, 1920), vii.

13. Donald L. Boese, *John C. Greenway and the Opening of the Western Mesabi* (Grand Rapids, MN: Department of History, Itasca Community College, 1975), 60.

14. Alfred Henry Lewis, ed., *A Compilation of the Messages and Speeches of Theodore Roosevelt, 1901–1905* (Washington, DC: Bureau of National Literature and Art, 1906), 677–78.

15. Roosevelt would reaffirm his tacit endorsement of mining later in life when he encouraged his sons Kermit and Theodore Jr. to consider careers in mine engineering, a fact documented in his correspondence with Greenway.

16. Interview, *Arkansas Gazette*, October 19, 1905.

17. Theodore Roosevelt to John Campbell Greenway, November 6, 1906, and January 30, 1907, John C. Greenway collection, Arizona Historical Society, Tucson.

18. Boese, *John C. Greenway and the Opening of the Western Mesabi*, 118–19.

19. Boese, *John C. Greenway and the Opening of the Western Mesabi*, 74–75.

20. "Theodore Roosevelt and Conservation," Theodore Roosevelt National Park, North Dakota, National Park Service, https://www.nps.gov/thro/learn/historyculture/theodore-roosevelt-and-conservation.htm; Gifford Pinchot, "The Conservation of Natural Resources" (1908), Digital Public Library of America, https://dp.la/primary-source-sets/environmental-preservation-in-the-progressive-era/sources/919.

21. Theodore Roosevelt, Seventh Annual Message, "To the Senate and House of Representatives," December 3, 1907. In *The Wilderness Warrior*, historian Douglas Brinkley notes, "More than any other policy Roosevelt adopted as president, the signing of the Antiquities Act has earned him praise from modern environmentalists[;] it represented the self-proclaimed 'wilderness hunter' as a high-minded naturalist statesman": Brinkley, *The Wilderness Warrior*.

22. Gifford Pinchot, *The Fight for Conservation* (1910), Project Gutenberg, https://www.gutenberg.org/files/11238/11238-h/11238-h.htm.

23. Kim Briggeman, "Words that Shaped Montana: 'Those Who Succeed Us,'" *Montana Magazine*, October 5, 2017; John Campbell Greenway to Theodore Roosevelt, April 8, 1906, Theodore Roosevelt Center, Dickinson University, Dickinson, ND.

24. Boese, *John C. Greenway and the Opening of the Western Mesabi*, 130–31.

25. Moving overburden also created Pill Hill in Hibbing, then the Winston and Dear Dump, referred to in chapter 1.

26. 190,602 acres for the Minnesota National Forest; 1,018,638 acres for the Superior National Forest. Not until decades later, through a series of presidential and legislative actions, were portions of the Superior National Forest preserved as wilderness areas.

27. Boese, *John C. Greenway and the Opening of the Western Mesabi*, 165.

28. *Skillings Mining Review* 29, no. 28 (November 2, 1940), cited at John Baeten, Ghost Plants blog, https://www.industriallandscapes.org/ghost-plants/canisteo-washing-plant-coleraine-mn; John Baeten, "Contamination as Artifact: Waste and the Presence of Absence at the Trout Lake Concentrator, Coleraine, Minnesota," in *Geographies of Post-Industrial Place, Memory, and Heritage* (Oxfordshire: Routledge, 2020).

29. Joel Cave, "John Campbell Greenway," manuscript, Itasca County Historical Society, Grand Rapids, MN.

30. In *Mining North America*, John R. McNeill writes, "Prior to the mid-twentieth century, miners and mining engineers often paid little attention to the ecological consequences of their trade. Reviewing the mining literature of the late nineteenth century, Duane Smith finds little mention of the environment among mining's proponents. When they did discuss the environment, it was only to emphasize its usefulness in mining operations. The few miners and engineers who spoke of the environment described the wilderness as a site for character building, but not preservation": John R. McNeill and George Vrtis, eds., *Mining North America: An Environmental History since 1522* (Oakland: University of California Press, 2017).

31. Baeten, "Contamination as Artifact," 87, 88.

32. John Baeten, "A Landscape of Water and Waste: Heritage Legacies and Environmental Change in the Mesabi Iron Range," PhD diss., Michigan Technological University, 2017, 193.

33. American Society of Civil Engineers, "Civil Engineers Create Wonders of the World," *Civil Engineering* (July/August 2021).

34. Marvin G. Lamppa, *Minnesota's Iron Country: Rich Ore, Rich Lives* (Duluth, MN: Lake Superior Port Cities, 2004), 174. The crowding effect between the towns and mines eventually resulted in the creation of the first commercial busing service on the Iron Range, which later grew into Greyhound Bus.

35. Edmond L. DeLestry, *Western Magazine* (May 1916).

36. Sigurd Olson, oral history interview, 1976, Ely, MN, 9, 15, Minnesota Historical Society, St. Paul; David Backes, *A Private Wilderness: The Journals of Sigurd F. Olson* (Minneapolis: University of Minnesota Press, 2021), 6.

37. Longyear, *Mesabi Pioneer*, 30; William Chalmers Agnew, "Recollections of the Early History of the Mahoning Ore and Steel Company Prior to 1909."

Notes to Chapter 3: The Pioneers

1. As a random example, contributor Rand Sturdy writes in the *Hometown Focus* newspaper, "The military material of the second world war came from this pit and others like it. Our cars, buildings, trains, tweezers, scalpel blades, vehicle springs and all the rest came from here": Rand Sturdy, "Open Your Eyes to the Beautiful Winter Landscape," *Hometown Focus*, January 29, 2022, 9. Manuel, *Taconite Dreams*, 46; "PM Producer," spring 1970, box 3, Oliver Iron Mining Company research files, Minnesota Historical Society, St. Paul.

   Syracuse Lake near Biwabik was drained to create the Embarrass Mine Pit in 1944. Locals still call it "Lake Mine" in memory of the lake that existed before the mine. Regarding Hull-Rust's massive output, recall my pit-side conversation with an awestruck German documented in chapter 1.

2. Ayn Rand, *Atlas Shrugged* (New York: Dutton, 2005), 372; Aaron Brown, "The Good Ship Taconite, Flagship of Empire Built on Mesabi Range Profits," Minnesota Brown: Modern Life in Northern Minnesota, May 7, 2017, https://minnesotabrown.com/2017/05/ship-taconite -mesabi-range-profits.html; Phil Davies, "Pedal to the Metal," Federal Reserve Bank of Minneapolis, July 18, 2014, https://www.minneapolis fed.org/article/2014/pedal-to-the-metal.

   Indeed, Hibbing's high school was a wonder, with the Belgian cut-glass chandeliers alone valued at over $2 million: John G. Krier, "Hibbing: The Richest Little Village" (Hibbing, MN: Hibbing Historical Society/Tribune Graphic Arts, 1981).

3. Gilbert A. Leisman, "A Vegetation and Soil Chronosequence on the Mesabi Iron Range Spoil Banks, Minnesota," *Ecological Monographs* 27, no. 3 (July 1957): 221–45.

4. Leisman, "A Vegetation and Soil Chronosequence," 223.

5. Leisman, "A Vegetation and Soil Chronosequence," 228.

6. The challenges aspen present in milling was learned from personal experience while trying to use a router to create a bullnose edge on aspen sauna benches and confirmed through conversation with a carpenter friend.

7. US Forest Service, "How Aspens Grow," https://www.fs.usda.gov/ wildflowers/beauty/aspen/grow.shtml.

8. Richard Powers, *The Overstory* (New York: W. W. Norton, 2019).

9. Leisman, "A Vegetation and Soil Chronosequence," 235.

10. Leisman, "A Vegetation and Soil Chronosequence," 237.

11. Cal Flyn, *Islands of Abandonment: Nature Rebounding in the Post-Human Landscape* (New York: Viking, 2021), 28, 30, 31.

12. Christa Lawler, "Bob Dylan's 'Girl from the North Country' (maybe) Dies in California: Echo Star Casey Was a Striking Beauty, the Brigitte Bardot of Hibbing High," (San Jose) *Mercury News*, January 23, 2018, https://www.mercurynews.com/2018/01/23/bob-dylan-girl-from-the-north-country-echo-star-casey-died-iron-range/.

13. Jon Blistein, "Bob Dylan Readies Iron Sculptures for Exhibit," *Rolling Stone*, September 24, 2013, https://www.rollingstone.com/music/music-news/bob-dylan-readies-iron-sculptures-for-exhibit-107267/; Anthony Scaduto, *The Dylan Tapes: Friends, Players and Lovers Talking Early Bob Dylan* (Minneapolis: University of Minnesota Press, 2022), 25–26; "Playboy Interview: Bob Dylan: A Candid Conversation with Iconoclastic Idol of the Folk-Rock Set," *Playboy* (February 1966).

14. Bob Dylan, "11 Outlined Epitaphs," 1963, 2; "Q&A with Bill Flanagan," March 22, 2017, the Official Bob Dylan Site, https://www.bobdylan.com/news/qa-with-bill-flanagan; Douglas Brinkley, "Bob Dylan's Late-Era, Old-Style Individualism," *Rolling Stone*, May 14, 2009.

15. Bob Dylan, *Biograph* (Columbia, 1986).

16. Ron Rosenbaum, "Bob Dylan: Playboy Interview," *Playboy* (March 1978).

17. "Remarks of Senator John F. Kennedy, Hibbing, Minnesota, October 2, 1960," John F. Kennedy Presidential Library and Museum, https://www.jfklibrary.org/archives/other-resources/john-f-kennedy-speeches/hibbing-mn-19601002.

18. Manuel, *Taconite Dreams*, 46.

19. Walter Havighurst, *Vein of Iron: The Pickands Mather Story* (Cleveland, OH: World Pub. Co., 1958), 204, 205.

20. Manuel, *Taconite Dreams*, 18.

21. Pamela Koch, *Taconite, New Life for Minnesota's Iron Range: The History of Erie Mining Company* (Brookfield, MO: Donning Co. Publishers, 2019), 93.

22. S. K. Dickinson and D. G. Youngman, "Taconite Tailing Basin Reclamation—A Phase of Multiple Resource Management," 32nd Annual Mining Symposium and 44th Annual Meeting of the Minnesota Section, American Institute of Mining, Metallurgical, and Petroleum Engineers, 1971.

23. Sam Dickinson, "Revegetation of Taconite Tailings," *Mining Congress Journal* (October 1972): 6.

24. Koch, *History of Erie Mining Company*, 106.

25. Thomas J. Manthey, "Mine Land Environment in Minnesota—A Progress Report," 1971, author's collection.

26. Daniel Philip Wiener, John DiStafano, Ron Lanoue, and Joseph Mohbat, *Reclaiming the West: The Coal Industry and Surface Mined Lands* (New York: Inform, Inc., 1980), 4.

27. David G. Youngman, "Mineland Reclamation at LTV Steel Mining Company: Learning from the Past, Preparing for the Future," *Achieving Land Use Potential through Reclamation: Proceedings of the 9th Annual National Meeting of the American Society for Surface Mining and Reclamation,* Duluth, Minnesota, June 14–18, 1992, 680.

28. Dickinson, "Revegetation of Taconite Tailings," 5.

Notes to Chapter 4: Rules and Agencies

1. In his first State of the State Address in 1977, Minnesota governor Rudy Perpich said that "only one state in the union ha[d] done less in mine-land reclamation than Minnesota" and that his administration "intend[ed] to improve our stand dramatically."

2. C. McKechnie Thomson and S. Rodin, "Colliery Spoil Tips: After Aberfan," https://www.icevirtuallibrary.com/doi/pdf/10.1680/iicep.1973.4702.

3. State of Minnesota, Office of Hearing Examiners for the Department of Natural Resources, DNR-80-001-HK, "Report of the Hearing Examiner in the Matter of the Proposed Rules Relating to Mineland Reclamation."

4. Gabriel Popkin, "The Green Miles," *Washington Post Magazine*, February 13, 2020.

5. Nonminer Deaths and Injuries in Open Mine Pits Related to Trespassing through Mine Pit Fencing, author's collection: see Appendix 1, page 209.

6. US Department of the Interior, "Surface Mining and Our Environment: A Special Report to the Nation" (Washington, DC: US Department of the Interior, 1967), 52.

7. Every year, tens of thousands of people visit the many "mine view" overlooks located on the Mesabi Range in Nashwauk, Hibbing, Mountain Iron, Virginia, and other places. Surely opinions vary about the landscape below these mine views. Even among my fellow Iron Range residents, I have heard everything from "The dumps are the best thing that has happened to this town" to "I have to move away to quit seeing them."

8. Several scram mining companies operated on the Mesabi Range from the 1980s through the 2010s, including Skubic Brothers, Rhude & Fryberger, and most recently Magnetation. As of this writing, no active

scram mining has been conducted since Magnetation declared bankruptcy in 2015.

9. *Iron Range Resources and Rehabilitation Board: Celebrating 75 Years on the Iron Range, 1941–2016* (Virginia, MN: W. A. Fisher Co., 2016). At the time this new taconite process was under development, new tax legislation, known as the Taconite Tax, was being written to enable it. The Taconite Tax bill changed the way iron ore was taxed from a property-tax type of system, which taxed the total value of a property (including the value of its iron ore, which is how the city of Hibbing had previously attained the title of "Richest Little Village in the World"), to a production-tax based system, in which ore was taxed when it was mined but left tax free for bust times when the economy made mining uneconomical. This change in taxation, as well as the depletion of the natural ore reserves and the industry's pivot to taconite mining, meant that local communities would generate far less mining tax revenue than they had in past decades.

10. Minnesota Department of Natural Resources, "Financial Assurance for Minnesota's Iron Mining Industry, A Report to the Legislature," December 1994.

11. Minutes of the Financial Assurance Advisory Committee Meeting, December 6, 1993, Natural Resource Research Institute, Duluth, MN.

12. Samuel Kent Dickinson, obituary, Rowe Funeral Home, Grand Rapids, MN, https://www.rowefuneralhomeandcrematory.com/obituary/2464073.

Notes to Chapter 5: Laurentian Vision

1. Greenway's title was superintendent of the Canisteo mine in Coleraine, whereas Swearingen's title was general manager for U.S. Steel's Minntac mine in Mountain Iron.

2. As an example: in 2020, the production tax was $2.856 per ton, resulting in $109,035,255 in tax revenue for the state of Minnesota.

3. Adelle Whitefoot, "Tom Rukavina Is Remembered for His Accomplishments and Big Personality," *Duluth News Tribune*, January 19, 2019, https://www.duluthnewstribune.com/news/tom-rukavina-is-remembered-for-his-accomplishments-and-big-personality.

4. "Progressive reclamation" in this sense means reclaiming some mined lands while other areas of active mining are still being conducted, as opposed to simply waiting until all mining is completed before reclamation activities are started.

5. "Pre-1980," meaning it had no formal reclamation requirements under the Minnesota Mineland Reclamation Rules.

6. Michael Tomberlin, "On the Record: Tom Howard of USS Real Estate Talks about Largest Landowner in County," AL.com, March 20, 2011, https://www.al.com/businessnews/2011/03/on_the_record_tom_howard_of_us.html.

7. James A. Stolpestad, *Great Northern Iron: James J. Hill's 109-Year Mining Trust* (St. Paul, MN: Ramsey County Historical Society, 2020), 192. As noted in previous chapters, the University of Minnesota has played a long and pivotal role in the development of the Mesabi Iron Range, from mineral processing research (such as E. W. Davis's work on the taconite process) to mineland reclamation (such as C. O. Rost's work on revegetation of taconite tailings) to the modern-day efforts of the Natural Resources Research Institute in Hermantown, Minnesota. The University of Minnesota also benefits directly from mining production taxes on school trust lands, estimated to be valued at over $12 million per year: Iron Mining Association of Minnesota, "Minnesota Iron Mining: Our Communities, State and Nation Depend on It," https://www.taconite.org/mining-industry.

In the interest of making the text more readable, the names of many LVP contributors have been omitted from this chapter. However, in order to recognize the importance of each and every contribution, the names and affiliations of the participants in the LVP's key meetings and public efforts are assembled in Appendix 2, page 211.

8. For more on the Merritt brothers and other Iron Range mining pioneers, see Paul D. Kruif, *Seven Iron Men: The Merritts and the Discovery of the Mesabi Range* (Minneapolis: University of Minnesota Press, 2007).

9. M. Christine Carlson, John Koepke, and Mirja P. Hanson, "From Pits and Piles to Lakes and Landscapes: Rebuilding Minnesota's Industrial Landscape using a Transdisciplinary Approach," *Landscape Journal* 30, no. 1 (2011): 35–52; Robert E. McHenry, *Chat Dumps of the Missouri Lead Belt, St. Francois County, with an Illustrated History of the Lead Companies that Built Them* (Adrian, MI: R. E. McHenry, 2008), ii; Flyn, *Islands of Abandonment*, 36.

10. University of Minnesota, "A Prospectus: The Laurentian Vision for the Iron Range of Minnesota," draft, November 18, 1998, courtesy of Dennis Hendricks, author's collection.

11. The Taconite Environmental Protection Fund was established by the 1977 legislature to reclaim, restore, and enhance areas of Minnesota that have been adversely affected by environmentally damaging operations involved in mining and producing taconite and iron ore concentrate. The scope of activities includes local economic development projects and mine reclamation on state-owned lands, as described in chapter 4.

12. Carlson, Koepke, and Hanson, "From Pits and Piles to Lakes and Land-scapes," 35–52.

13. As stated in the land acknowledgment earlier in the book, the name *Iron Range* refers primarily to white settlement and naming of the region, which is much different from the names and land uses of the Indigenous populations that occupied the Mesabi area prior to iron mining.

14. Carlson, Koepke, and Hanson, "From Pits and Piles to Lakes and Land-scapes," 37.

15. Irina A. Fursman, "How Leadership Occurs in a Loosely Coupled, Multi-Stakeholder System," PhD diss., University of St. Thomas, 2021, https://ir.stthomas.edu/celc_ed_old_conf/2/.

16. University of Minnesota, "The Laurentian Vision for the Iron Range of Minnesota."

17. Brian Hiti, "The Laurentian Vision: A Process to Enhance the Future of the Iron Range," draft, 2000, Iron Range Resources and Rehabilitation Board, courtesy of Dennis Hendricks, author's collection.

18. Christine Carlson and John Koepke, "The Laurentian Vision: Rebuild-ing Minnesota's Mining Region," *SCAPE: Land and Design in the Upper Midwest* 17, no. 25 (2017).

19. Carlson, Koepke, and Hanson, "From Pits and Piles to Lakes and Land-scapes," 35.

20. LVP Coordination Meeting 12, Meeting Report, February 8, 2001, cour-tesy of Dennis Hendricks, author's collection.

21. Definitions from Merriam-Webster and Oxford English Dictionary.

22. Stolpestad, *Great Northern Iron*, 197.

23. Carlson, Koepke, and Hanson, "From Pits and Piles to Lakes and Land-scapes," 36, 47.

24. Carlson, Koepke, and Hanson, "From Pits and Piles to Lakes and Land-scapes," 42; grant application to Blandin Foundation, July 26, 2002, U.S. Steel archives.

25. Carlson, Koepke, and Hanson, "From Pits and Piles to Lakes and Land-scapes," 45.

26. Carlson, Koepke, and Hanson, "From Pits and Piles to Lakes and Land-scapes," 45.

27. John Koepke and Chris Carlson, "The Laurentian Vision Partnership: Transforming Pits and Piles into Lakes and Landscapes," presentation for the Minnesota Shade Tree Short Course, University of Minnesota, Department of Landscape Architecture, March 2018.

28. Carlson, Koepke, and Hanson, "From Pits and Piles to Lakes and Landscapes,"46.

29. As listed in Stolpestad, *Great Northern Iron*, 195.
30. Mineland Vision Partnership, "Strategic Plan 2020–2025," https://mvpmn.org/wp-content/uploads/2022/04/MVP-2020_2025-Strategic-Plan-Final-w-Appendix.pdf.
31. Minutes of Financial Assurance Advisory Committee, December 6, 1993, Natural Resource Research Institute, Duluth, at which the question was asked, "How many acres are still out there that are not covered by the reclamation rules?" It was estimated that there was one hundred years' worth of mining before reclamation, so there is one hundred years' worth of work remaining. An inventory in 1977 showed approximately 130,000 acres were disturbed by mining. Presently, approximately 150,000 acres have been disturbed since mining began. The Permit to Mine areas cover about 200,000 acres.

    The iron ore resources on the Mesabi remain vast, especially if "down-dip" ores located beneath the Virginia slate are included. If the price of iron ore suddenly rose by ten or twenty times, many locations that are currently considered uneconomical to mine might come into contention for future mining.
32. Carlson, Koepke, and Hanson, "From Pits and Piles to Lakes and Landscapes," 36.
33. This built feature—which is easy to directly compare to the LVP's Gateway Bridge—is why the Virginia charrette drawing was selected as a focus of this book. However, the Virginia, Hibbing-Chisholm, and Biwabik charrette drawings all contain thoughtful, visually captivating, and inspirational images and concerns that could be used to guide future mineland reclamation and community revitalization efforts.
34. Creating outcomes beyond the minimum requirements of the Mineland Reclamation Rules.
35. Whereas the same was never true of Greenway, Cole, Longyear, Hibbing, Chisholm, and others who died far from the parks and schools and municipalities that bear their names.

Notes to Chapter 6: The Future Today

1. In "No Direction Home," Bob Dylan said the Mesabi Range was simply too cold for acting out. Perhaps for this reason, Hibbing was once named the "Safest Town in America": Steve Tanko, "What Is the Safest City in the United States? Hibbing Minnesota Tops the List," May 4, 2015, Kool 101.7.
2. North Star Destination Strategies, Hibbing Brand Presentation, 2007, author's collection.

3. "ZipUSA: 55746," *National Geographic* (October 2002).

4. Flyn, *Islands of Abandonment*, 32, 33.

5. Robin Dean, "From Dereliction to Inspiration: How Romania's Mine Remediation Program Is Transforming Lives," *Engineering and Mining Journal* 212, no. 7 (September 2011): 64–65.

6. This LVP charrette drawing is no longer held by the local library, having been culled from the collection because of low usage.

7. Plummer's remark came despite the fact that I never mentioned nor aspired for public office. My described naivete stemmed from the fact that at the time I didn't know about any of the mine reclamation and repurposing work described in the previous chapters or the work in progress of the Laurentian Vision Partnership.

8. "Everything You Need to Know about Travel in 2021," *Outside Magazine*, February 1, 2021, https://www.outsideonline.com/2420541/ everything-you-need-know-about-travel-2021.

9. The resistance to repurposing had by then attracted criticism by some citizens in the form of letters to the editor. Around the same time, I recall the city administrator refusing match-free grant money on the speculation that "it might someday cost us a dollar."

10. Olga Smirnova, "Sisu: The Finnish Art of Inner Strength," Worklife, BBC, May 7, 2018, https://www.bbc.com/worklife/article/20180502 -sisu-the-finnish-art-of-inner-strength.

11. Minnesota Statute 180.03, Subpart 4.a.

12. Baeten, "A Landscape of Water and Waste: Heritage, Legacies and Environmental Change in the Mesabi Iron Range," 39, citing Walter L. Thurman, "Waste Dumps of the Mesabi Iron Range: Heritage or Blight?," MA thesis, St. Cloud State University, 1992; Elisabeth Deffner, "It's All Downhill," American Profile, December 24, 2000, http:// americanprofile.com/articles/ski-jumping-michigan-town; Paul Mitchell, "Top 10 Business Risks and Opportunities for Mining and Metals in 2021," Ernst & Young, October 7, 2021, https://www.ey.com/ en_gl/mining-metals/top-10-business-risks-and-opportunities-for -mining-and-metals-in-2022.

13. Jeff Schoenbauer, "Findings and Recommendations: Redhead Mountain Bike Trail Facility," June 29, 2017, author's collection.

14. The work and business owned and operated by Adam Harju and his wife, Micah, was later highlighted by the National Public Radio program *Marketplace* on October 7, 2021.

15. Minnesota Pollution Control Agency, "Environmental Justice," https:// www.pca.state.mn.us/about-mpca/mpca-and-environmental-justice.

16. A provision that was made good in 2021 when a grant-in-aid ATV trail through the park was finalized.

17. Rudy Davison and John S. Pillsbury III, "The Pillsbury Family Connection to Iron Ore Mining in Northern Minnesota from 1888 to the Present," *Mining History Journal* 17 (2010): 26–43.

18. Megan Michelson, "9 New U.S. Trails You Should Try This Year," *Outside*, April 19, 2021, https://www.outsideonline.com/adventure-travel/ destinations/north-america/new-american-trails-2021/; Conor Mihell, "Shredding as a Path to Community Development," *Sierra*, May 2, 2021, https://www.sierraclub.org/sierra/shredding-path-community -development.

19. Marie Tolonen, "Bringing New Life to Hibbing's Historic Reed Building," *Mesabi Tribune*, February 13, 2022, https://www.mesabitribune .com/news/local/bringing-new-life-to-hibbing-s-historic-reed -building/article_ef57d226-8c63-11ec-939e-5ba425b27abf.html.

20. *Reclaimed*, Redhead Mountain Bike Park, 13:14, https://www.youtube. com/watch?v=Qpscdce31YM; *Biketown: How Trails Transform Communities* short film, *Freehub Magazine*, 39:51, https://www.youtube.com/ watch?v=bBcFzjP31jQ. As a basis of comparison, the Magnetation projects referenced earlier in the chapter had ended in bankruptcy, devastating several established businesses and families on the Iron Range, and the Essar Steel project was never built to completion. Therefore, while Redhead's economic contribution might be small, it was at least adding to the economy of the Iron Range during a period when no new mining companies had been able to succeed.

21. Those "full-scale manifestations" do not include the original demonstration projects.

    This chapter documents and honors the contributions of people who made Redhead happen, including Jim Plummer, Dan Jordan, Tony Sertich, and Mark Phillips; the board and the staff of the IRRR; Mark Casey, Mike Jugovich, Marty Halverson, Mary Jo Rahja, and other elected officials and staff of the city of Chisholm; Jim Shoberg; the Minnesota Discovery Center; Jeff Schoenbauer; the St. Louis County mine inspectors and leaders; Barr Engineering Co.; the iron mining industry; the MDNR; Kay-Linn; the talented cadre of trail builders, the volunteers of IROC, and the citizens of Chisholm and the Iron Range. Today, the Redhead Mountain Bike Park project is being led by a new crew of volunteers and professionals who are lending some of those "hands on the wheel." The leadership from the three partners who own and operate the park includes Mayor John Champa, councilor Travis Vake, and city administrator Stephanie Skraba from the city of Chisholm; Donna Johnson and Jordan Metsa from the Minnesota Discovery Center; Gary Sjoquist; volunteer dirt bosses Pat Cassingham and Robert Crowe; and the newly elected volunteer leaders for IROC, Kari Kilen, Melissa Crowe, Benji Neff, Briana Sterle, and Miriam Kero. While this list names and enumerates the individuals and entities who built Redhead and keep it going today, it also demonstrates another, perhaps unexpected, benefit of the project: its ability to build community.

Notes to Epilogue

1. Lee Bloomquist, "Iron Range Endowed with More Than Enough Ore," *Mesabi Daily News*, October 30, 2019. For example, Butler Taconite closed in 1985 with enough ore remaining in the ground that a whole other mining operation—Essar Steel Minnesota Limited, whose groundbreaking in 2007 is described in chapter 6—planned on several more decades of mining before going bankrupt with its plant only half constructed.
2. Maya Wei-Haas, "See Relics of Europe's Industrial Past Reimagined as Amusement Parks," *National Geographic*, photos and video by Luca Locatelli, November 30, 2022.
3. See brand survey results in chapter 6. David M. Abbott Jr., "Accepting the Depletability of Deposits and the Creation of Ghost Towns," *Mining Engineering* 73, no. 11 (November 2021): 33.
4. DeLestry, "The Story of Hibbing, Wonder-Village of the World."
5. Thomas Michael Power, *The Economic Role of Metal Mining in Minnesota: Past, Present and Future*, Minnesota Center for Environmental Advocacy and the Sierra Club, October 2007, 7, 26, 27, 28, 31.
6. Mihell, "Shredding as a Path to Community Development"; Heather Hansman, "Why We Should Be Turning Former Mines into Trails," *Outdoor*, May 29, 2021, https://www.outsideonline.com/outdoor -adventure/environment/why-we-should-be-turning-former-mines -trails.
7. Power, *The Economic Role of Metal Mining in Minnesota*, 7.
8. The final cost was $230 million, with $30 million coming from the federal government and the remaining from the state: "Thomas Rukavina Memorial Bridge," Wikipedia, https://en.wikipedia.org/wiki/Thomas_ Rukavina_Memorial_Bridge. Redhead Mountain Bike Park cost $1.777 million to build and maintain to date, funded by the IRRR.

Index

Page numbers in *italic* refer to illustrations.

Minescapes has been typeset in Adobe Text Pro, typeface family designed by Robert Slimbach. Adobe Text bridges the gap between calligraphic Renaissance types of the fifteenth and sixteenth centuries and high-contrast Modern styles of the eighteenth century, taking many of its design cues from early post-Renaissance Baroque transitional types cut by designers such as Christoffel van Dijck, Miklós (Nicholas) Kis, and William Caslon.

Interior text design by Wendy Holdman